SETTING PRI for LARGE RESEARCH FACILITY PROJECTS supported by the NATIONAL SCIENCE FOUNDATION

Committee on Setting Priorities for
NSF-Sponsored Large Research Facility Projects

Committee on Science, Engineering, and Public Policy
Policy and Global Affairs Division

Board on Physics and Astronomy
Division on Engineering and Physical Sciences

THE NATIONAL ACADEMIES

THE NATIONAL ACADEMIES PRESS
Washington, D.C.
www.nap.edu

THE NATIONAL ACADEMIES PRESS 500 Fifth Street, N.W. Washington, DC 20001

NOTICE: The project that is the subject of this report was approved by the Governing Board of the National Research Council, whose members are drawn from the councils of the National Academy of Sciences, the National Academy of Engineering, and the Institute of Medicine. The members of the committee responsible for the report were chosen for their special competences and with regard for appropriate balance.

Support for this project was provided by the National Science Foundation (under grant OIA-0304912). Any opinions, findings, conclusions, or recommendations expressed in this publication are those of the author(s) and do not necessarily reflect the views of the organizations or agencies that provided support for the project.

International Standard Book Number 0-309-09084-9 (Book)
International Standard Book Number 0-309-52775-9 (PDF)

Available from the Committee on Science, Engineering, and Public Policy; and the Board on Physics and Astronomy, 500 Fifth Street, NW, Washington, DC 20001; 202-334-2807; Internet, http://www.nationalacademies.org/cosepup and http://www.nationalacademies.org/bpa. Additional copies of this report are available from the National Academies Press, 500 Fifth Street, NW, Lockbox 285, Washington, DC 20055; (800) 624-6242 or (202) 334-3313 (in the Washington metropolitan area); Internet, http://www.nap.edu.

THE NATIONAL ACADEMIES
Advisers to the Nation on Science, Engineering, and Medicine

The **National Academy of Sciences** is a private, nonprofit, self-perpetuating society of distinguished scholars engaged in scientific and engineering research, dedicated to the furtherance of science and technology and to their use for the general welfare. Upon the authority of the charter granted to it by the Congress in 1863, the Academy has a mandate that requires it to advise the federal government on scientific and technical matters. Dr. Bruce M. Alberts is president of the National Academy of Sciences.

The **National Academy of Engineering** was established in 1964, under the charter of the National Academy of Sciences, as a parallel organization of outstanding engineers. It is autonomous in its administration and in the selection of its members, sharing with the National Academy of Sciences the responsibility for advising the federal government. The National Academy of Engineering also sponsors engineering programs aimed at meeting national needs, encourages education and research, and recognizes the superior achievements of engineers. Dr. Wm. A. Wulf is president of the National Academy of Engineering.

The **Institute of Medicine** was established in 1970 by the National Academy of Sciences to secure the services of eminent members of appropriate professions in the examination of policy matters pertaining to the health of the public. The Institute acts under the responsibility given to the National Academy of Sciences by its congressional charter to be an adviser to the federal government and, upon its own initiative, to identify issues of medical care, research, and education. Dr. Harvey V. Fineberg is president of the Institute of Medicine.

The **National Research Council** was organized by the National Academy of Sciences in 1916 to associate the broad community of science and technology with the Academy's purposes of furthering knowledge and advising the federal government. Functioning in accordance with general policies determined by the Academy, the Council has become the principal operating agency of both the National Academy of Sciences and the National Academy of Engineering in providing services to the government, the public, and the scientific and engineering communities. The Council is administered jointly by both Academies and the Institute of Medicine. Dr. Bruce M. Alberts and Dr. Wm. A. Wulf are chair and vice chair, respectively, of the National Research Council.

www.national-academies.org

Preface

For many years, policy makers and the scientific community have focused attention on the support provided by the National Science Foundation (NSF) for large facilities used in scientific and engineering research. Previous reports have addressed the complex issues that arise in choosing among facility proposals and in balancing support for facilities and other tools with support for research conducted by individual investigators.[1] As large facilities have become an increasingly prominent part of the nation's research and development portfolio and as NSF has entered a period of budgetary expansion, concerns once again have intensified.

In a letter to the president of the National Academies dated June 12, 2002, Senators Barbara Mikulski, Christopher Bond, Ernest Hollings, John

---

[1] Appendix G provides the executive summary of the Committee on Science, Engineering, and Public Policy's previous report from the Panel on NSF Decisionmaking for Major Awards, chaired by Robert Rutford, titled *Major Award Decisionmaking at the National Science Foundation* (Washington, D.C.: National Academy Press, 1994). Other reports of interest include reports from the National Science Board, *Criteria for the Selection of Research Projects by the National Science Foundation* (Washington, D.C.: National Science Foundation, 1974); Office of Technology Assessment, US Congress, *Federally Funded Research: Decisions for a Decade* (Washington, D.C.: US Government Printing Office, 1991); President's Council of Advisors on Science and Technology, *Megaprojects in the Sciences* (Washington, D.C.: Office of Science and Technology Policy, Executive Office of the President, 1992); and National Science Board, *Science and Engineering Infrastructure for the 21st Century: The Role of the National Science Foundation* (Arlington, Virginia: National Science Foundation, 2003).

McCain, Edward Kennedy, and Judd Gregg stated that "questions remain as to whether the NSF has a satisfactory process for prioritizing multiple competing large-scale research facility projects." The letter said that NSF funding of requests for large facility projects appears to be "ad hoc and subjective." It also pointed out that the NSF inspector general had recently found "significant deficiencies in the Foundation's management and oversight of its large facility projects resulting in significant cost overruns not contemplated in their original budgets." To address those concerns— which also have been expressed by members of the House Committee on Science and by the members and staffs of other congressional committees and subcommittees—the letter requested that the National Academy of Sciences "review the current prioritization process and report to us on how it can be improved."

In response to the request, the National Academies appointed the Committee on Setting Priorities for NSF-Sponsored Large Research Facility Projects[2] to address the following charge:

- Review NSF's current prioritization process as well as processes and procedures used by other relevant organizations.
- Develop the criteria that should be considered in developing priorities among competing large research facility proposals.
- Provide recommendations for optimizing and strengthening the process used by the NSF to set priorities among large research facility project proposals and to manage their incorporation into the President's budget.
- Provide recommendations for improving the construction and operation of NSF-funded large research facility projects.
- Provide recommendations regarding the role of the current and future availability of international and interagency research facility projects in the decision-making process for NSF funding of large research facility projects.[3]

This report focuses on a portion of NSF's activities that is small (less than 4 perccent) compared with the foundation's overall budget but is nevertheless central to its mission. It examines the policies and procedures governing awards made through the Major Research Equipment and Facilities Construction (MREFC) account.[4] NSF uses the MREFC

---

[2]Appendix A provides biographical information on the committee members.

[3]Appendix B provides a copy of the charge, the Senate letter, and related congressional documents.

[4]Appendix C provides histories of all current MREFC projects and those approved but not funded.

account to support the "acquisition, construction, commissioning, and upgrading of major research equipment, facilities, and other such capital assets" that cost more than several tens of millions of dollars. The report looks at how plans and proposals for large research facilities originate, how NSF chooses which facilities to support, and how it oversees their construction. These "large research facility projects" represent major investments in the future of a given field of research. Funding the construction of a large facility affects the direction of research for many years and implies continued support for the operations and maintenance of the facility. Large research facilities also can have a substantial effect on regional economies, public perceptions of science, workforce training, and international cooperation in research. NSF's support of large facility projects is a critical element of US science and technology policy and warrants sustained attention from policy makers and the research community.

In responding to its charge, the committee examined numerous NSF documents,[5] National Science Board (NSB) minutes and presentations, congressional testimony, and news articles, web sites, and reports that discuss the facilities. The committee also compared NSF's current process with that used by the Department of Energy (DOE) Office of Science, the National Aeronautics and Space Administration (NASA) Office of Space Science, the United Kingdom, and Germany.[6] In addition, it compiled examples of criteria that have been used or proposed for use to set research priorities by various organizations and several countries.[7] This study also builds on the Committee on Science, Engineering, and Public Policy's (COSEPUP's) 1994 report *Major Award Decisionmaking at NSF*, which addressed some of the same issues that are of concern here.[8] Finally, the committee had useful discussions with the staff of the Senate Appropriations Committee, the House Science Committee, NSF, DOE, NASA, and disciplinary societies and researchers. Those people are listed in the acknowledgments section.

Given the ever-changing and draft nature of NSF's process for setting priorities among its proposals for large research facilities, the committee decided that it would not be fair to NSF to conduct an investigation of each decision it had made since 1995 (when the MREFC account was created) or even earlier (when some of the current projects began construction). The committee chose instead to examine the process as it exists today as outlined by NSF and to focus on how that process can be improved from the time of project conception to operation.

---

[5]See key excerpts in Appendix F.
[6]See Appendix D.
[7]See Appendix E.
[8]See Appendix G.

In doing so, the committee concluded that although NSF has improved its process for setting priorities among large facility projects, further strengthening is needed, if NSF is to meet the demands that will be made of it in the future. This report lays out specific recommendations that describe how large facility projects should be ranked within and among disciplines. In addition, it discusses how NSF can enhance preapproval planning and budgeting of projects and oversight of construction and operation once projects are approved to ensure that the nation's investment is ultimately successful. As research opportunities and agency initiatives change, the recommendations in this report should remain at the core of the procedures used to identify, develop, set priorities among, and manage large facility projects. By implementing the report's recommendations, NSF, in partnership with the research community, can develop a system of short-term and long-term planning that is sufficiently robust to direct funding to the most meritorious research projects. In that way, NSF can increase its already substantial contributions to the nation's science and engineering enterprise.

William F. Brinkman, *Chair*
Committee on Setting Priorities for NSF-Sponsored Large Research Facility Projects

Acknowledgments

This report is the product of many people. First, we thank all those who provided information at our committee meetings in 2003. They were (in alphabetical order)

MARC ALLEN, Director, Strategic and International Planning, National Aeronautics and Space Administration

BARRY BARISH, Maxine and Ronald Linde Professor of Physics and Director of Laser Interferometer Gravitational Wave Observatory Laboratory, California Institute of Technology

JOSEPH BORDOGNA, Deputy Director, National Science Foundation

RITA COLWELL, Director, National Science Foundation

PATRICIA DEHMER, Associate Director, Office of Basic Energy Sciences, Office of Science, Department of Energy

ADRIENNE FROELICH, Director of Public Policy, American Institute of Biological Sciences

SCOTT GILES, Deputy Chief of Staff, Majority Staff, House Committee on Science

DAVID GOLDSTON, Chief of Staff, Majority Staff, House Committee on Science

MARTHA HAYNES, Goldwin Smith Professor of Astronomy and Director of Undergraduate Studies, Cornell University

ANITA JONES, Chair, Committee on Programs and Plans, National Science Board

CHEH KIM, Professional Staff Member, Majority Staff, Senate Appropriations Committee

MIKE LUBELL, Director of Public Affairs, American Physical Society, and Chairman, Physics Department, City College of New York

JOHN MARBURGER III, Director, White House Office of Science and Technology Policy

KEVIN MARVEL, Deputy Executive Officer, American Astronomical Society

TED MOORE, Professor of Geological Sciences, University of Michigan

ERIC NAGY, President, Organization of the Biological Field Stations, and Associate Director, Mountain Lake Biological Station, and Assistant Professor, Department of Biology, University of Virginia

RAY ORBACH, Director, Office of Science, Department of Energy

DAVID RADZANOWSKI, Branch Chief, Science and Space Programs Branch, Office of Management and Budget

PETER ROONEY, Staff Director, Majority Staff, Research Subcommittee, House Committee on Science

JAMES TIEDJE, University Distinguished Professor of Microbiology and Director of National Science Foundation Center for Microbial Ecology, Michigan State University

DAVID TRINKLE, Program Examiner, Science and Space Programs Branch, Office of Management and Budget

JIM WILSON, Professional Staff Member, Minority Staff, Research Subcommittee, House Committee on Science

Without the input of each of those speakers, this report would not have been possible.

This report has been reviewed in draft form by individuals chosen for their diverse perspectives and technical expertise, in accordance with procedures approved by the NRC's Report Review Committee. The purpose of the independent review is to provide candid and critical comments that will assist the institution in making the published report as sound as possible and to ensure that the report meets institutional standards for objectivity, evidence, and responsiveness to the study charge. The review comments and draft manuscript remain confidential to protect the integrity of the deliberative process.

We wish to thank the following for their participation in the review of this report: John Armstrong (Retired), IBM Corp; Arthur Bienenstock, Stanford University; Bernard Burke, Massachusetts Institute of Technology; Claude Canizares, Massachusetts Institute of Technology; James Clark, Duke University; John Evans (Retired), Comcast Corporation; Paul Gaffney, Monmouth University; Norine Noonan, College of Charleston; Yves Petroff, European Synchrotron Radiation Facility; J. Michael Rowe,

National Institute of Standards and Technology; and Patrick Windham, Windham Consulting.

Although the reviewers listed above have provided many constructive comments and suggestions, they were not asked to endorse the conclusions or recommendations, nor did they see the final draft of the report before its release. The review of this report was overseen by John Ahearne, Sigma Xi (The Scientific Research Society), and France Cordova, University of California, Riverside. Appointed by the National Research Council, they were responsible for making certain that an independent examination of this report was carried out in accordance with institutional procedures and that all review comments were carefully considered. Responsibility for the final content of this report rests entirely with the author committee and the institution.

In addition, we thank the guidance group that oversaw this project:

MAXINE SINGER (*Chair*), President Emerita, Carnegie Institution of Washington
SHIRLEY CHIANG, Professor of Physics, Department of Physics, University of California, Davis
JAMES DUDERSTADT, President Emeritus and University Professor of Science and Engineering, University of Michigan
ELSA GARMIRE, Sydney E. Junkins Professor of Engineering, Dartmouth College
JOHN HUCHRA, Professor, Harvard-Smithsonian Center for Astrophysics
BURTON RICHTER, Paul Pigott Professor in the Physical Sciences, Stanford University, and Director Emeritus, Stanford Linear Accelerator Center
HUGO SONNENSCHEIN, Charles L. Hutchinson Distinguished Service Professor, Department of Economics, University of Chicago

Finally, we thank the staff of this project, including Deborah Stine, associate director of the Committee on Science, Engineering, and Public Policy and study director, who managed the project; Timothy Meyer, program officer of the Board on Physics and Astronomy; Steve Olson, the science writer for this report; Kevin Rowan and Laura Holliday, who provided project support; National Academies Science and Technology Policy Interns Jonathan Tucker, who conducted the initial background research for the committee, and Blake Purnell and Arti Garg, who developed the historical analysis in Appendix C; Donald Shapero, director of the Board on Physics and Astronomy; and Richard Bissell, executive director of the Committee on Science, Engineering, and Public Policy and of the Policy and Global Affairs Division.

Contents

APPENDIXES

SETTING PRIORITIES
for LARGE RESEARCH
FACILITY PROJECTS
supported by the
NATIONAL SCIENCE
FOUNDATION

Executive Summary

L arge facilities play a more prominent role in science and engineering research today than they have played in the past. In FY 1995, the National Science Foundation (NSF) created the Major Research Equipment and Facilities Construction (MREFC) account to support the "acquisition, construction, commissioning, and upgrading of major research equipment, facilities, and other such capital assets" that cost more than several tens of millions of dollars.

Although such large facility projects represent less than 4 percent of the total NSF budget, they are highly visible because of their large per-project budget, their potential to shape the course of future research, the economic benefits they bring to particular regions, and the prominence of the facilities in an increasing number of research fields.

A number of concerns have been expressed by policy makers and researchers about the process used to rank large research facility projects for funding. First, the ability of new projects to be considered for approval at the National Science Board (NSB) level has stalled in the face of a backlog of approved but unfunded projects. Second, the rationale and criteria used to select projects and set priorities among projects for MREFC funding have not been clearly and publicly articulated. Third, there is a lack of funding for disciplines to conduct idea-generating and project-ranking activities and, once ideas have some level of approval, a lack of funding for conceptual development, planning, engineering, and design—information needed when judging whether a project is ready for funding in light of its ranking and for preparing a project for funding if it is selected. Those concerns have eroded confidence among policy makers and the research community that large research facility projects are being ranked on the basis of their potential returns to science, technology, and society.

To address the concerns regarding NSF's process for identifying, approving, constructing, and managing large research facility projects, the committee makes the following recommendations:

1. **The National Science Board should oversee a process whereby the National Science Foundation produces a roadmap for large research facility projects that it is considering for construction over the next 10-20 years.**

Broad inputs from the scientific community must form the basis for the roadmap.

The roadmap should take into consideration the need for continued funding of existing projects and should provide a set of well-defined potential new project starts for the near term (0-10 years). These projects should be ranked against other projects expected to be funded in a given year and according to where they are positioned in time on the roadmap. Projects further out in time (10-20 years) will necessarily be less well defined and ranked qualitatively to yield a vision of the future rather than a precise funding agenda, as is the case for the earlier years.

Different categories of overlapping criteria, described briefly in the bullets below, need to be used as one moves from comparing projects within a field to comparing projects in a directorate or in the entire NSF. At each level, the criteria used in the previous level must continue to be considered.

- Within a field (as defined by NSF division) or interdisciplinary area: *scientific and technical criteria,* such as scientific breakthrough potential and technological readiness.
- Across a set of related fields: *agency strategic criteria,* such as balance across fields and opportunities to serve researchers in several disciplines.
- Across all fields: *national criteria* that assess relative need—such as which projects maintain US leadership in key scientific and engineering fields or enable the greatest numbers of researchers, educators, and students.

See box on page 4 for a more in-depth discussion of the proposed criteria.

A key constraint that must be imposed in the final stages of development is that the roadmap must reflect a reasonable projection of the large research facility budget over the next 2 decades. The roadmap is not a guarantee of funding but rather a plan for the development of NSF's large research facility program.

2. The National Science Foundation, with the approval of the National Science Board, should base its annual MREFC budget submission to Congress on the roadmap. The annual budget submission should include the proposed yearly expenditures over the next 5 years for committed projects and for projects that will start in that period. It should supply a rank ordering of the proposed new starts and should include the rationale behind the proposed budget, the project ranking, and any differences between the budget submission and the roadmap.

The committee emphasizes that the final determination and approval of rankings across disciplines must be the responsibility of the NSF senior leadership subject to final approval by the NSB.

3. To ensure that a large research facility project selected for funding is executed properly, on schedule, and within its budget, the National Science Foundation should enhance project preapproval planning and budgeting to develop a clear understanding of the project's "technical definition" (also called "scope of work") and the "implementation plan" needed to carry out the work.

Once a project is funded, there should be provision for a disciplined periodic independent review of the project's progress relative to the original plan by a committee that includes internal and external engineering and construction experts and scientific experts and that will monitor the project's status and provide its evaluation to the NSB and NSF.

After the construction phase, a committee with a different external and internal membership that includes scientists and people with expertise in managing large facilities should monitor facility operations annually (or as needed).

Finally, NSF has created a new position—Deputy Director, Large Facility Projects in NSF's Office of Budget, Finance, and Award Management—to oversee the construction of these projects. Given the new nature and importance of this position, it should be reviewed by a committee of internal and external experts to evaluate its operation and effectiveness within a 2-year period. (See page 17 for a description of this position.)

4. To ensure that potential international and interagency collaborations and ideas are discussed at the earliest possible stages, the Office of Science and Technology Policy in the Executive Office of the President

Criteria for Developing Large Facilities Roadmaps and Budgets

Overlapping categories of criteria should guide the preparation of the large facilities roadmap and NSF's annual budget submissions. Scientific and technical quality must be at the core of these criteria. Because these are large facility projects, they must have the potential to have a major impact on the science involved; otherwise, they should not reach the next step.

The rankings show what we would expect to happen first within a field, then within a directorate of NSF, and then across NSF. The criteria from earlier stages must continue to be used as the ranking proceeds from one stage to the next.

- First Ranking: Scientific and Technical Criteria Assessed by Researchers in a Field or Interdisciplinary Area
  - Which projects have the most scientific merit, potential, and opportunities within a field or interdisciplinary area?
  - Which projects are the most technologically ready?
  - Are the scientific credentials of the proposers of the highest rank?
  - Are the project-management capabilities of the proposal team of the highest quality?

- Second Ranking: Agency Strategic Criteria Assessed Across Related Fields by Using the Advice of Directorate Advisory Committees
  - Which projects will have the greatest impact on scientific advances in this set of related fields taking into account the importance of balance among fields for NSF's portfolio management in the nation's interest?
  - Which projects include opportunities to serve the needs of researchers from multiple disciplines or the ability to facilitate interdisciplinary research?
  - Which projects have major commitments from other agencies or countries that should be considered?
  - Which projects have the greatest potential for education and workforce development?
  - Which projects have the most readiness for further development and construction?

should have a substantial early role in coordinating roadmaps across agencies and with other countries.

5. Given the congressional emphasis in the most recent National Science Foundation reauthorization bill and the importance of the MREFC account to the research community and the health of the US research enterprise, the NSF leadership and the NSB must give careful attention to the implementation of reforms in the MREFC account.

- **Third Ranking: National Criteria Assessed Across All Fields by the National Science Board**
  - Which projects are in new and emerging fields that have the most potential to be transformative? Which projects have the most potential to change how research is conducted or to expand fundamental science and engineering frontiers?
  - Which projects have the greatest potential for maintaining US leadership in key science and engineering fields?
  - Which projects produce the greatest benefits in numbers of researchers, educators, and students enabled?
  - Which projects most need to be undertaken in the near term? Which ones have the most current windows of opportunity, pressing needs, and international or interagency commitments that must be met?
  - Which projects will have the greatest impact on current national priorities and needs?
  - Which projects have the greatest degree of community support?
  - Which projects will have the greatest impact on scientific advances across fields taking into account the importance of balance among fields for NSF's portfolio management in the nation's interest?

Ranking projects across disciplines is inherently not an exact science; nevertheless, these criteria, as illustrated by the questions, provide a framework for a discussion of why one project is accorded a higher priority than another and a mechanism for the discussion to be as objective as possible in ranking projects across fields.

Within the ranking categories, the questions might change as governmentwide initiatives and unexpected occurrences shift priorities. Similarly, at times, some questions might have greater weight than others in the judgment of the NSB. The key element is for the questions and weighting to be identified before the ranking process begins and for a clear rationalization to be provided when proposed large research facility projects are ranked.

The committee has outlined a six-step process to implement these recommendations.

Large research facility projects will continue to constitute a vital component of NSF's science and technology portfolio by enabling researchers to examine previously inaccessible phenomena and answer previously intractable questions. NSF has strengthened the priority-setting process for these facilities in recent years, partly in response to reports from Congress and other organizations.

NSF now has an opportunity to strengthen the program further by incorporating the preparation of a roadmap into its planning process and by involving the research community more fully in the generation and ranking of ideas for large research facilities.

Making choices among competing proposals from different scientific fields will never be easy, but the recommendations and detailed steps described here can help NSF to excel in this critical part of its mission.

Introduction

The National Academies Committee on Setting Priorities for National Science Foundation (NSF)-sponsored Large Research Facility Projects was charged with examining NSF's current process for setting priorities, developing criteria that should be considered in setting priorities, and providing recommendations as to how that process can be strengthened and optimized. The committee was also to provide recommendations regarding the role that the availability of interagency and international large facility projects should play in the decision-making process. Furthermore, the committee was to provide recommendations as to how construction and operation of these facilities can be improved.

The large facility projects supported by NSF are nearly as varied as the scientific research that the foundation supports. Some facilities include new and increasingly powerful versions of instruments that have been used for decades to study the natural world, such as telescopes or particle accelerators. Other large facilities use new ways of gathering information; examples are a new facility designed to measure gravity waves generated by such cosmic events as star collisions and supernovae and a proposed facility that would detect high-energy neutrinos in a large volume of Antarctic ice to provide information about the astrophysical sources of extremely high-energy cosmic rays. Some large facilities primarily serve specific scientific disciplines, such as optical telescopes and radio-telescopes for astronomy and observatory networks for oceanography. Other facilities enable research in a wide array of disciplines; for example, the ground facilities, ships, and aircraft stationed in Antarctica allow scientists to study the atmosphere, ice, oceans, and geology of the region.

Regardless of their detailed characteristics, all large facility projects are being affected by the accelerating development of information technologies. Increasing quantities and varieties of information are being gathered, rapidly analyzed, and interpreted. Information technologies are also changing the fundamental nature of many large facility projects. New information technologies are making it possible, for example, for many large facilities to consist of smaller instruments and research projects in widely distributed geographic locations. The George E. Brown, Jr. Network for Earthquake Engineering Simulation, which is intended to improve the seismic design and performance of the US civil and mechanical infrastructure, will consist of 15 experimental equipment sites linked by a high-performance Internet system. Elements of EarthScope, a distributed project to study the structure and dynamics of North America, will operate in nearly every county in the United States during the project's lifetime of 8-10 years. The proposed National Ecological Observatory Network would consist of geographically distributed observatories linked to laboratories, data archives, and computer modeling facilities.

In FY 1995, NSF created the Major Research Equipment and Facilities Construction (MREFC) account to support the "acquisition, construction, commissioning, and upgrading of major research equipment, facilities, and other such capital assets" that cost more than several tens of millions of dollars.

The MREFC account was created to separate the construction funding for a large facility—which can rise and fall dramatically over the course of a few years—from the more continuous funding of facility operations and individual-investigator research. The account, however, has remained too small to fund all the large facility projects that NSF would like to undertake now and in the future, as the National Science Board (NSB) points out in its report *Science and Engineering Infrastructure for the 21st Century.*

Of NSF's FY 2004 budget request of $5.48 billion, about 24.5 percent, or $1.34 billion, is for the budget category of "tools," which includes support for large facilities and the small- and medium-scale infrastructure needed for state-of-the-art research.[1] The remainder of NSF's research and education budget is divided into two additional categories: "ideas" and "people." In the tools category, the request for the MREFC account was $202 million—about 15 percent of the tools budget request and about 3.7 percent of NSF's total request.

---

[1]National Science Foundation, *FY 2004 Budget Request.*

Despite representing less than 4 percent of the total NSF budget, the facility projects supported through the MREFC account are highly visible because of their large project budgets, their potential to shape the course of future research in one or more fields, their potential economic benefits for particular regions, their effects on international cooperation in research, and their prominence in an increasing number of research fields. They also represent initial investments in particular fields that will require continuing support to operate, maintain, and perhaps upgrade. In addition, many of the issues raised by these projects must also be considered in terms of their impact on other NSF projects and programs as NSF proposes its portfolio of investments for each fiscal year.

Description of
National Science Foundation's
Current Process

T he following sections provide a description of NSF's current process for its large research facilities from the time an idea originates through concept development, priority-setting, implementation, and oversight of its construction and operation.

ORIGINS OF CONCEPT AND DEVELOPMENT OF PROPOSALS[1]

The origins of large facility projects are as varied as the projects themselves. Some arise as logical outgrowths of previous research or facilities. Others originate as a consequence of new scientific development when the need for a new facility becomes apparent where no such need existed before. For example, high-speed networks and computers enable data acquisition and processing over widely dispersed geographic areas, creating the need for new large integration facilities.

The impetus for all new large facility projects originates in the scientific community, but ideas take various routes to fruition. The community processes vary greatly from field to field. Often, self-organizing groups within a field of science or engineering develop the initial ideas for a new facility and set scientific objectives for the facility by ranking

---

[1]To ensure that this description of NSF's current process was accurate, given the evolving nature of NSF's priority-setting procedures for the MREFC account, the committee decided to send the draft text to NSF for review. NSF reviewed, revised, and approved the portions of the text that reference this footnote.

competing needs. At other times, facilities have been proposed at the initiative of an individual scientist or a small group of researchers with a bold vision. NSF program officers and staff foster these initiatives by providing funds for meetings and workshops that facilitate the scientific community's internal evaluation and maturation of these concepts. In every case, the mission of NSF is to seek out the best ideas and the best scientists and to empower their investigations.

This process of nurturing and maturation of a concept for a facility can take many years to develop fully or it can come together as a funded proposal quite quickly, depending on the nature of the proposal, the immediacy of the scientific need, and the potential payoffs scientifically and for society in general. NSF's role in this process is reactive and responsive to the scientific community, rather than prescriptive, and this ensures that the highest quality proposals, as determined by peer review within the scientific community, are brought forward for implementation. NSF program officers are the key people who make the requirements for approval of such projects clear to the community.

In identifying new facility construction projects, the science and engineering community, in consultation with NSF, develops ideas, considers alternatives, explores partnerships, and develops cost and timeline estimates. By the time a proposal is submitted to NSF, those issues have been thoroughly examined.

ESTABLISHING PRIORITIES FOR LARGE FACILITY PROJECTS[2]

On receipt by NSF, large facility proposals are first subjected to rigorous external peer review that focuses on the criteria of intellectual merit and broad (probable) impacts. Only the highest rated proposals—those rated outstanding on both criteria—survive this process. These are recommended for further review by an MREFC panel that comprises the NSF assistant directors and office heads, who serve as stewards for their fields and are chosen for their breadth of understanding, and is chaired by the NSF deputy director acting in consultation with the director and later for review by the NSB.

Both the MREFC panel and the NSB look for a consistent set of attributes in each project that they recommend:

• The project represents an exceptional opportunity to enable frontier research and education.

---

[2]See footnote 1.

- The impact on a particular field of research is expected to be transformational.
- The relevant research community places a high priority on the project.
- The resulting facility will be accessible to an appropriately broad user community.
- Partnership possibilities for development and operation are fully exploited.
- The project is technically feasible, and potential risks are thoroughly addressed.
- There is a high state of readiness—with respect to engineering cost effectiveness, interagency and international partnerships, and management—to proceed with development.

The MREFC review panel evaluates the merit of a proposed project and then ranks it against other projects under consideration. It first selects the new projects that it will recommend to the director for future NSF support on the basis of a discussion of the merits of the science in the context of all sciences that NSF supports. Projects that are not highly rated according to the above criteria are returned to the initiating directorates and may be reconsidered later. Highly rated projects are placed in priority order by the panel in consultation with the NSF director. The review panel and the director emphasize the following criteria to determine the priority order of the projects:

- How "transformative" is the project? Will it change how research is conducted or alter fundamental science and engineering concepts or research frontiers?
- How great are the benefits of the project? How many researchers, educators, and students will it enable? Does it broadly serve many disciplines?
- How pressing is the need? Is there a window of opportunity? Are there interagency and international commitments that must be met?

Those criteria are not assigned relative weights, because each project has its own unique attributes and circumstances. For example, timeliness may be crucial for one project and relatively unimportant for another. And the director must weigh the impact of a proposed facility on the balance between scientific fields, the importance of the project with respect to national priorities, and possible societal benefits.

NSF DIRECTOR AND NATIONAL SCIENCE BOARD[3]

Using the recommendations received from the MREFC panel, the NSF director selects candidate projects to be considered by the NSB during one of its meetings. According to the *Guidelines*, the director uses the following criteria in making this selection:

- Strength and substance of the information provided to the MREFC panel.
- The relationship to NSF goals and priorities, including NSF's educational mission.
- Appropriate balance among various fields, disciplines, and directorates on the basis of consideration of needs and opportunities.
- Guidance from the NSB on overall decision boundaries for the MREFC account provided at the annual MREFC planning discussion (May).
- Opportunities to leverage NSF funds.

The NSB's Committee on Program and Plans (CPP) takes the lead in reviewing a proposed project; a member of the committee leads the discussion. The CPP uses these criteria:
- Need for such a facility.
- Research that will be enabled.
- Readiness of plans for construction and operation.
- Construction budget estimates.
- Operations budget estimates.

After the CPP reviews the project, it makes recommendations to NSB for approving its inclusion in future budget requests and for approving project implementation.

NSF DIRECTOR AND OFFICE OF MANAGEMENT AND BUDGET

Once the NSB has approved a project for funding, the director may recommend the project for inclusion in a future budget request to the Office of Management and Budget (OMB).

In August of each year, the director presents the priorities, including a discussion of the rationale for the priority order, to the NSB as part of the budget process. The NSB reviews the list and either approves or argues the order of priority.

---

[3]The text of this section has been reviewed and approved by the NSB.

As part of its budget submission, NSF presents the rank-ordered list of projects to OMB. For projects included in the budget request, a capital asset plan and justification must be prepared; it follows a format developed by OMB. The capital asset plan summarizes how much the project will cost to build and operate, information on its management and cost, its schedule, and performance goals and milestones.

The list of major projects in the budget may be modified during negotiations between OMB and NSF. During that process, other parts of the executive branch, such as the White House Office of Science and Technology Policy, may provide input on the projects included in the budget. Finally, NSF submits a priority list of projects to Congress as part of its budget submission.

CONGRESSIONAL ACTION

After submission of the president's budget to Congress in February, congressional subcommittees and committees examine the proposed expenditures and begin the appropriations process. Congressional appropriators make decisions about whether to fund each of the large facility projects proposed for NSF in the president's budget. In addition, because of budgetary constraints, the NSF director and OMB may decide not to request funds for large facility projects that the NSB has approved for inclusion in the budget.

By 2001, the NSB had approved six large facility projects that had not yet been funded. Concerns were expressed in Congress and elsewhere that political pressures rather than scientific merit would increasingly determine which projects received appropriations. In 2001, Congress asked NSF to rank the six projects in order of priority. NSF responded by dividing the projects into two categories of three projects each; there was no ordering within a category. In its appropriations for FY 2003, Congress provided funds for two of the three projects in the high-priority category. In the 2004 budget request, NSF further ranked the projects, requesting funding for the remaining high-priority project in that fiscal year and proposing to start funding for the other three in FY 2005 and FY 2006.

PROJECT IMPLEMENTATION AND OVERSIGHT[4]

Except for its facilities in the Antarctic, NSF does not directly operate research facilities. Rather, it makes awards to other organizations—such as universities, consortia of universities, and nonprofit organizations—to

---

[4]See footnote 1.

construct, operate, and manage the facilities. NSF enters into partnerships with those organizations, whose details are most often defined through cooperative agreements, to accomplish that. A cooperative agreement defines the scope of work to be undertaken by an awardee and establishes the project-specific terms and conditions by which NSF will maintain oversight of the project. NSF has the final responsibility for oversight of the development, management, and performance of a facility.

Each large facility project supported by NSF has a program manager in NSF who is the primary person responsible for all aspects of project oversight and management of the project within the foundation. The program manager carries out these responsibilities in accordance with an internal management plan (IMP) that has been crafted specifically for the project. The IMP defines a project advisory team (PAT) that consists of NSF personnel with expertise in the scientific, technical, management, and administrative issues associated with the project. The team works with the program manager to ensure the establishment of realistic cost, schedule, and performance goals for the project. The team also helps to develop terms and conditions of awards for constructing, acquiring, and operating a large facility. NSF's deputy for large facility projects works closely with the program manager, providing expert assistance on non-scientific and nontechnical aspects of project planning, budgeting, implementation, and management to strengthen the oversight capabilities of the foundation. The deputy also facilitates the use of best management practices by fostering coordination and collaboration throughout NSF to share application of lessons learned from prior projects.

The awardee designates one person to be the project director. This person has overall control and responsibility for the project within the awardee organization. Throughout the implementation stage, the awardee executes and manages the project—either construction or acquisition—in accordance with the cooperative agreement between the awardee institution and NSF. This phase of the project includes all installation, testing, commissioning, and acceptance. Oversight by NSF during this phase is accomplished through periodic reviews, written reports by the awardee to the foundation that include documentation of technical and financial status based on "earned value" reporting methods, annual work plans, periodic external reviews, and site visits.

By the end of the implementation stage, a proposal is submitted to the program manager for operations and maintenance. The program manager reviews proposals in accordance with the merit-review procedures in Chapter V of NSF's *Proposal and Award Manual* and presents a recommendation for funding to his or her division director and assistant director-office head. The director's review board (DRB) reviews proposals for awards that exceed the DRB threshold (provided in Chapter VI of NSF's

Proposal and Award Manual). After DRB review, the NSF director recommends awards above the NSB threshold to the NSB for approval. The NSB reviews and approves awards recommended by the director. The assistant director-office head, through the division director, then authorizes the program manager to recommend the making of an award in accordance with the proposal-processing procedures contained in Chapter VI of the *Proposal and Award Manual.*

The program manager, with the Division of Grants and Agreements, drafts the cooperative agreement that will govern the project in accordance with the procedures contained in Chapter VIII of the *Proposal and Award Manual.* The Division of Grants and Agreements makes the award once the cooperative agreement is executed by it and the awardee.

DEPUTY DIRECTOR, LARGE FACILITY PROJECTS, OFFICE OF BUDGET, FINANCE, AND AWARD MANAGEMENT[5]

As part of its *Large Facility Projects Management and Oversight Plan* drafted in 2001, NSF created a new position—deputy director, large facility projects—in the Office of Budget, Finance, and Award Management—to enable consistent management and oversight of all large projects, including all business and financial aspects, from their conceptualization phase through operations. The first person to serve in this capacity assumed office in June 2003. This official, who reports to the director of the Office of Budget, Finance, and Award Management (who is also NSF's chief financial officer), is responsible for

• Providing expert assistance to NSF's science and engineering staff on project planning, budgeting, implementation, and management.
• Developing, implementing, and managing—with NSF-wide input and concurrence—management and oversight policies, guidelines, and procedures.
• Ensuring shared learning of best practices by fostering coordination and collaboration throughout NSF to facilitate application of lessons learned from each project.
• Chairing and convening the NSF Facilities Panel, which establishes the appropriate level of management and oversight NSF will apply to each large facility project.
• Monitoring the business operations aspects of the facilities.
• Ensuring consistent representation of NSF staff on project advisory teams that advise and assist program officers in charge of large facility

---
[5]See footnote 1.

projects in establishing realistic cost, schedule, and performance goals; developing the terms and conditions of cooperative agreements; overseeing projects; and providing assistance in moving projects through exceptional situations that may arise.

This official is consulted on all policy matters involving facilities, including responses to inquiries made by NSF management, the NSF Office of Inspector General, the Office of Management and Budget, and Congress. The position is supported by permanent NSF staff with a mix of skills, qualifications, and extensive experience in project management, planning and budgeting, cost analysis, and oversight. NSF refers to this new position as the "deputy for large facility projects," nomenclature that the committee respects throughout the rest of this report.

Concerns About the National Science Foundation's Current Priority-Setting Process

The recent focus on NSF's setting of priorities among large facility projects continues a long-running discussion of the best way for NSF to support such undertakings. In the June 12, 2002, letter to National Academy of Sciences President Bruce Alberts that led to the present study, six senators stated that "funding requests by the Foundation for large facility projects appear to be ad hoc and subjective." The letter directed the National Academies to "review the current prioritization process and report to us on how it can be improved."

In responding to this charge, the committee found that a number of concerns have been expressed by policy makers and researchers about the process used to rank large research facility projects for funding. First, the ability of new projects to be considered for approval at the NSB level has stalled in the face of a backlog of approved but unfunded projects. Second, the rationale and criteria used to select and set priorities among projects for MREFC funding have not been clearly and publicly articulated. Third, there is a lack of funding for disciplines to conduct idea-generation and, once ideas have some level of approval, there is a lack of funding for conceptual development, planning, engineering, and design—information needed to judge adequately whether a project is ready for full funding. Those concerns have eroded confidence among policy makers and the research community that large research facility projects are being ranked on the basis of their potential returns to science, technology, and society.

Large research facility projects have become too complex, expensive, and numerous to handle with procedures that may have sufficed in the past. NSF has improved the administration of the MREFC program over the last decade, but further improvements are necessary and possible.

In response to the first concern, in its 2003 report *Science and Engineering Infrastructure for the 21st Century: The Role of the National Science Foundation*, the NSB examined the status of the nation's science and engineering infrastructure and concluded that

> there is an urgent need to increase Federal investments to provide access for scientists and engineers to the latest and best S&E infrastructure, as well as to update infrastructure currently in place.

The NSB recommended that NSF "increase the share of the NSF budget devoted to S&E infrastructure," and one subject of emphasis was large facility projects.

In its report, the NSB noted that large facility project needs identified 5-10 years ago have not yet been funded although the scientific justification for the facilities has grown. Although the FY 2003 appropriation for the MREFC account was $148 million, an annual investment of $350 million for several years would be needed to address the backlog of projects.

As an example of the second concern, although the FY 2004 budget request did set priorities among projects, there was no justification for giving one project a higher priority than another as requested by Congress. It is only in the followup correspondence between NSF and the House Science Committee (see Appendix F) that the reasons become clearer. The lack of transparency has eroded confidence among policy makers and the research community and increased concerns that projects are not being chosen solely on the basis of merit. This lack of confidence in NSF's priority-setting process has increased the risk that large facility projects will be funded for reasons other than their potential returns to science, technology, and society.

NSF's support of large research facilities has undergone important improvements since the MREFC account was established in FY 1995. For example, NSF has institutionalized and publicized procedures governing the preparation and review of proposals.[1] However, the committee has received a number of comments regarding the lack of funds for the development of new projects. NSF should consider a more adequate process for ensuring the financing of such activity.

---

[1]National Science Foundation, *Facilities Management and Oversight Guide* (Arlington, Virginia: National Science Foundation, draft, November 8, 2002).

Recommendations

To address the concerns regarding NSF's current process for identifying, approving, constructing, and managing large research facility projects, the committee offers the following recommendations:

1. The National Science Board should oversee a process whereby the National Science Foundation produces a roadmap for large research facility projects that it is considering for construction over the next 10-20 years.

Broad inputs from the scientific community must form the basis for the roadmap.

The roadmap should take into consideration the need for continued funding of existing projects and should provide a set of well-defined potential new project starts for the near term (0-10 years). These projects should be ranked against other projects expected to be funded in a given year and according to where they are positioned in time on the roadmap. Projects further out in time (10-20 years) will necessarily be less well defined and ranked qualitatively, to yield a vision of the future rather than a precise funding agenda, as is the case for the earlier years.

Different categories of overlapping criteria, described briefly in the bullets below, need to be used as one moves from comparing projects within a field to comparing projects in a directorate or in the entire NSF. At each level, the criteria used in the previous level must continue to be considered.

- Within a field (as defined by NSF division) or interdisciplinary area: *scientific and technical criteria,* such as scientific breakthrough potential and technological readiness.
- Across a set of related fields: *agency strategic criteria,* such as balance across fields and opportunities to serve researchers in several disciplines.
- Across all fields: *national criteria* that assess relative need—such as which projects maintain US leadership in key scientific and engineering fields or enable the greatest numbers of researchers, educators, and students.

See box below for a more in-depth discussion of the proposed criteria.

Criteria for Developing Large Facilities Roadmaps and Budgets

Overlapping categories of criteria should guide the preparation of the large facilities roadmap and NSF's annual budget submissions. As shown in Figure 1, scientific and technical quality must be at the core of these criteria. Because these are large facility projects, they must have the potential to have a major impact on the science involved; otherwise, they should not reach the next step.

The rankings show what we would expect to happen first within a field, then within a directorate of NSF, and then across NSF. The criteria from earlier stages must continue to be used as the ranking proceeds from one stage to the next.

- **First Ranking: Scientific and Technical Criteria Assessed by Researchers in a Field or Interdisciplinary Area**
  - Which projects have the most scientific merit, potential, and opportunities within a field or interdisciplinary area?
  - Which projects are the most technologically ready?
  - Are the scientific credentials of the proposers of the highest rank?
  - Are the project-management capabilities of the proposal team of the highest quality?

- **Second Ranking: Agency Strategic Criteria Assessed Across Related Fields by Using the Advice of Directorate Advisory Committees**
  - Which projects will have the greatest impact on scientific advances in this set of related fields taking into account the importance of balance among fields for NSF's portfolio management in the nation's interest?
  - Which projects include opportunities to serve the needs of researchers from multiple disciplines or the ability to facilitate interdisciplinary research?
  - Which projects have major commitments from other agencies or countries that should be considered?
  - Which projects have the greatest potential for education and workforce development?
  - Which projects have the most readiness for further development and construction?

A key constraint that must be imposed in the final stages of development is that the roadmap must reflect a reasonable projection of the large research facility budget over the next 2 decades. The roadmap is not a guarantee of funding but rather a plan for the development of NSF's large research facility program.

Clearly, no one can project budgets out 20 years. However, one can expect to have rough estimates of the cost of a project that would allow NSF to plan for the future and provide guidance for future planning and design seed money. Thus, to create a credible roadmap, NSF would construct a tentative budget that might look something like the schematic in Figure 2. It would probably not be published with explicit budget

---

- **Third Ranking:** National Criteria Assessed Across All Fields by the National Science Board
  - ○ Which projects are in new and emerging fields that have the most potential to be transformative? Which projects have the most potential to change how research is conducted or to expand fundamental science and engineering frontiers?
  - ○ Which projects have the greatest potential for maintaining US leadership in key science and engineering fields?
  - ○ Which projects produce the greatest benefits in numbers of researchers, educators, and students enabled?
  - ○ Which projects most need to be undertaken in the near term? Which ones have the most current windows of opportunity, pressing needs, and international or interagency commitments that must be met?
  - ○ Which projects will have the greatest impact on current national priorities and needs?
  - ○ Which projects have the greatest degree of community support?
  - ○ Which projects will have the greatest impact on scientific advances across fields taking into account the importance of balance among fields for NSF's portfolio management in the nation's interest?

Ranking projects across disciplines is inherently not an exact science; nevertheless, these criteria, as illustrated by the questions, provide a framework for a discussion of why one project is accorded a higher priority than another and a mechanism for the discussion to be as objective as possible in ranking projects across fields.

Within the ranking categories, the questions might change as governmentwide initiatives and unexpected occurrences shift priorities. Similarly, at times, some questions might have greater weight than others in the judgment of the NSB. The key element is for the questions and weighting to be identified before the ranking process begins and for a clear rationalization to be provided when proposed large research facility projects are ranked.

National Criteria

Agency Strategic Criteria

Scientific and Technical Criteria

- Which projects have the most scientific merit, potential, and opportunities within a field or interdisciplinary area?
- Which projects are the most technologically ready?
- Are the scientific credentials of the proposers of the highest rank?
- Are the project management capabilities of the proposal team of the highest quality?

- Which projects will have the greatest impact on scientific advances in this set of related fields, taking into account the importance of balance among fields for NSF's portfolio management in the nation's interest?
- Which project include opportunities to serve the needs of researchers from multiple disciplines or the ability to facilitate interdisciplinary research?
- Which projects have major commitments from other agencies or countries that should be considered?
- Which projects have the greatest potential for education and workforce development?
- Which projects have the most readiness for further development and construction?

- Which projects are in new and emerging fields that have the most potential to be transformative? Which projects have the most potential to change how research is conducted or to expand fundamental science and engineering frontiers?
- Which projects have the greatest potential for maintaining US leadership in key scientific and engineering fields?
- Which projects produce the greatest benefits in numbers of researchers, educators, and students enabled?
- Which projects most need to be undertaken in the near term? Which ones have the most current windows of opportunity, pressing needs, and international or interagency commitments that must be met?
- Which projects will have the greatest impact on current national priorities and needs?
- Which projects have the greatest degree of community support?
- Which projects will have the greatest impact on scientific advances across fields, taking into account the importance of balance among fields for NSF's portfolio management in the nation's interest?

FIGURE 1 Projects are compared and ranked at different steps of the priority-setting process; at each step, additional categories of criteria should be applied. The innermost category of *scientific and technical criteria* remains at the core of the entire process; it is first applied to projects within the same field. As projects develop and are compared across divisions within the directorate, *agency strategic criteria* must also be considered. Finally, projects must be compared across all fields at NSF, and *national criteria* must be considered for final selection and ranking. The overlapping ellipses in this diagram emphasize the "transcend and include" aspect of the model: at each step, the criteria used at the previous step must continue to be considered.

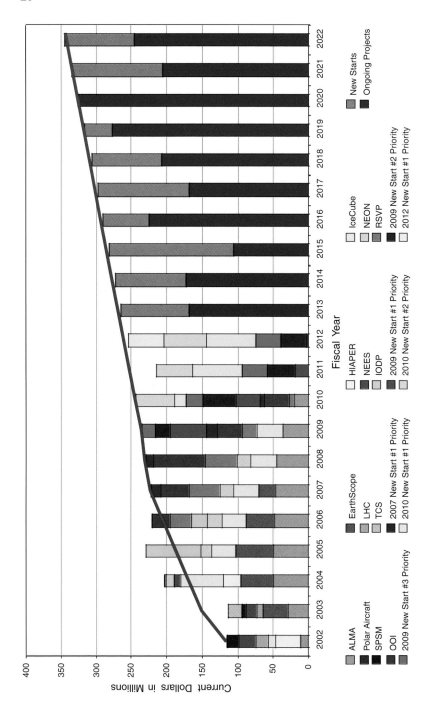

FIGURE 2 An example of a 20-year budget roadmap for large facility projects. This figure is intended to illustrate some of the temporal and fiscal considerations that should inform the NSF roadmapping process.

The MREFC funding amounts depicted for existing (already funded or planned) individual projects through FY 2008 reflect the FY 2004 budget submission. The MREFC funding levels shown for existing projects beyond FY 2008 are taken from the individual funding profiles of each project.

Through FY 2007, the budget envelope shown in this figure reflects the amount of MREFC funding indicated in the 2002 NSF Doubling Act. After this doubling, an annual growth rate of 3 percent was assumed.

The first 10 years depicted in this figure demonstrate the detailed, project-specific budget information that the near-term roadmap should provide. Further into the future, when plans are less definite, the figure indicates, roughly, the amount of funding available in a given year for all new starts and for all current MREFC account projects without providing detailed information about individual projects.

The particulars of the long-term projections will necessarily change over time, but such a 20-year plan will provide a realistic vision of possibilities for future new starts. While including this figure, the committee recognizes that it may be challenging to publish a long-term budget and believes that there are other ways to represent this information (see, for example, the DOE roadmap report).

numbers, inasmuch as the numbers for projects further out on the road-map will have to be approximate, and it is not good policy to preempt the president and Congress on budgets far out in the future. But that does not preclude using budget estimates as guidance.

NSF and the NSB should engage the scientific and engineering research community more widely in generating new ideas and initial proposals for large research facility projects and in providing comments directly throughout the approval process. They should make every effort to hear directly from the proposers of the projects.

The NSB has a particularly important role to play as the independent voice of the research community in ensuring that NSF's process is trans-parent and responds to the community and reforms proposed to the process.

In developing proposals for large research facilities, the potential for interagency and international collaboration should be explored at an ear-lier stage with idea generation, and the ranking process should take into account the plans of other agencies and countries.

In some research fields, particularly fields that overlap or fall between disciplines, NSF staff should more actively foster the planning and assess-ment of large research facility projects. A broad and balanced portfolio of projects—disciplinary and interdisciplinary—needs to emerge from the research community.

The research community should identify and rank projects at an early stage at the discipline and related-discipline level. Through activities funded by NSF, the research community should be involved in ranking project ideas within individual fields. In some fields—such as astronomy, astrophysics, and high-energy physics—this planning already occurs. NSF should rank proposed projects across the fields encompassed in each of its directorates by using the advice of its directorate advisory commit-tees. Ranking of projects above the directorate level should be proposed to the NSF director by a panel of top-level NSF officials. The NSF director will then propose the roadmap to the NSB, which by law has approval authority over all large facility projects.

Setting priorities across fields of science is a complex undertaking, but the committee believes that this responsibility belongs squarely with the senior leadership of NSF working with the NSB. At the same time, NSF and the NSB should make every effort to solicit input from the scien-tific community in producing the roadmap, for example, by placing a draft of the roadmap on NSF's web site for public comment. All com-ments on the roadmap and NSF's responses to the comments could also be placed on the web site. The resulting NSF roadmap should be widely distributed to the research community.

The final approval of the roadmap must lie with the NSB. It should work closely with the NSF staff and director to ensure the highest quality of roadmap. The inclusion of a project on the 20-year roadmap would not necessarily guarantee funding in the president's annual budget requests, in that priorities can change. However, the near-term projects should be closely reflected in the budget; if not, there should be an explanation. Also, depending on the state of affairs when NSF carries out the road-mapping process, one possible outcome is that the projects currently back-logged may not be considered to be the top priority at the time the roadmapping is conducted.

In the course of its work, the committee received a number of recom-mendations that it consider as models for the roadmapping processes. One example is that used by the high-energy physics community under the direction of the Department of Energy. Another is the roadmap developed by the astronomy community under the aegis of the National Aeronautics and Space Administration. We have adopted that idea, but we also hasten to point out that the process will have to be modified to mesh with NSF's distinctive culture, budgets, and missions and, particu-larly, to encourage the involvement of scientists in fields that have tradi-tionally not organized themselves in this fashion. Success will depend not only on NSF's efforts but also on the good will of the members of the research community asked to think across disciplinary lines in the course of advising NSF on the creation of the roadmap.

2. The National Science Foundation, with the approval of the National Science Board, should base its annual MREFC budget sub-mission to Congress on the roadmap. The annual budget submission should include the proposed yearly expenditures over the next 5 years for committed projects and for projects that will start in that period. It should supply a rank ordering of the proposed new starts and should include the rationale behind the proposed budget, the project ranking, and any differences between the budget submission and the roadmap.

The committee emphasizes that the final determination and approval of rankings across disciplines must be the responsibility of the NSF senior leadership subject to final approval by the NSB.

The observations and rationale used to rank one large facility project idea over another for inclusion in NSF's annual budget requests should be clearly and publicly described so that policy makers and researchers understand the motivation for the decisions. NSF's FY 2004 budget request and its followup letter to Congress are initial steps in that direction, but NSF should expand its discussion to respond to the need. The rationale should be prepared every year and should accompany, rather than follow, NSF's annual budget requests.

Inevitably, the budgetary framework assumed in developing the roadmap will not reflect actual annual budgets, so deviations from the roadmap necessary to prepare annual budget requests should be based on the above criteria, and the rationale for the decisions (or for deviation from the criteria) should be provided in the budget. Figure 3 summarizes this process by depicting the sequence of selection and ranking steps through which a large facility project must pass before being included in the president's budget request as a candidate for MREFC funding.

3. To ensure that a large research facility project selected for funding is executed properly, on schedule, and within its budget, the National Science Foundation should enhance project preapproval planning and budgeting to develop a clear understanding of the project's "technical definition" (also called "scope of work") and the "implementation plan" needed to carry out the work.

Once a project is funded, there should be provision for a disciplined periodic independent review of the project's progress relative to the original plan by a committee that includes internal and external engineering and construction experts and scientific experts and that will monitor the project's status and provide its evaluation to the NSB and NSF.

After the construction phase, a committee with a different external and internal membership that includes scientists and people with expertise in managing large facilities should monitor facility operations annually (or as needed).

Finally, NSF has created a new position—Deputy Director, Large Facility Projects in NSF's Office of Budget, Finance, and Award Management—to oversee the construction of these projects. Given the new nature and importance of this position, it should be reviewed by a committee of internal and external experts to evaluate its operation and effectiveness within a 2-year period.

If a project is highly ranked at the directorate level and is a candidate for a proposed start over the following 5 years, conceptual or preliminary engineering and plans for carrying out the project, in addition to the information described by the criteria, should be prepared before the NSB approves it for inclusion in the roadmap. The conceptual plans developed as a result of this process should describe how the project will be constructed and managed and should describe any constraints on its scope and implementation, such as funding, technology, development, or siting.

The conceptual plans should include the "technical definition" to provide sufficient information on the scope of the project. In complex projects, minor changes can evolve as more detail is developed, but there should be no major changes from this point out, without a thorough analysis that weighs the potential advantage of such changes against the impact on the project's cost and schedule and a review of that analysis by the National Science Board before initiation.

The "implementation plan" describes how the construction project will be accomplished. It includes such items as organization for implementation, management, contracting, scheduling, and budgeting. In addition, there should be plans and provisions for effective cost and schedule control, approval for changes, monitoring of progress, periodic review, and so on, all the way to commissioning and turning over the facility to the ultimate "owner and operator."

As mentioned above, the roadmap should be the guide to funding project planning, engineering, and design activities to prepare a project proposal for MREFC funding. NSF has frequently been criticized for not supplying funding for this phase of a project.

During the construction of a large research facility, review committees should consist primarily of experts in constructing new and unique facilities and experts in the scientific and technical subjects that the project entails.

While the facility is in operation, the review committees should include people who have experience in managing large research facilities and, again, experts in the scientific and technical fields that use the facility. The value of informed, independent, and objective review in managing project construction and operations cannot be overemphasized.

The role of the deputy for large facility projects is to manage this process and bring all the various constituencies together so that the project happens on time, within budget, and with satisfactory performance. As a result, the person needs adequate and experienced project construction and management staff, access to qualified consultants and contractors, and the institutional authority to oversee the design engineering, construction, and operation phases adequately. Each project or program will have dedicated leadership, but it is this deputy who has principal responsibility to support the undertakings and for oversight and management. Because this deputy plays such a critical role, the office should be reviewed within 2 years to ensure that it is adequately staffed and providing the appropriate level of project oversight and leadership.

4. To ensure that potential international and interagency collaborations and ideas are discussed at the earliest possible stages, the Office of Science and Technology Policy in the Executive Office of the President should have a substantial early role in coordinating roadmaps across agencies and with other countries.

As noted in the remarks about the first recommendation, early discussions regarding potential interagency and international coordination

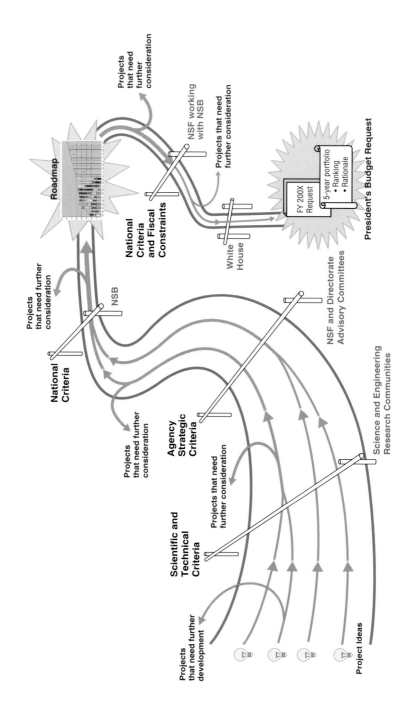

Projects that need further consideration

Roadmap

NSF working with NSB

Projects that need further consideration

National Criteria and Fiscal Constraints

White House

FY 200X Request

5-year portfolio
• Ranking
• Rationale

President's Budget Request

Projects that need further consideration

NSB

National Criteria

Projects that need further consideration

Agency Strategic Criteria

NSF and Directorate Advisory Committees

Scientific and Technical Criteria

Projects that need further consideration

Science and Engineering Research Communities

Projects that need further development

Project Ideas

FIGURE 3 Depiction of the sequence of selection and ranking steps through which a large facility project must pass before being included in the president's budget request as a candidate for MREFC funding. Each gateway represents a phase in the priority-setting process of selection and ranking with oversight by a specific body. The category of criteria identified at each gateway is detailed in the box on page 22. Note that international and interagency cooperation considerations are explicit items in both the *agency strategic criteria* and *national criteria* categories and therefore are important throughout the entire process.

and collaboration can reduce the financial costs of a project to participating agencies and countries and can enhance its prospects of success.

5. Given the congressional emphasis in the most recent National Science Foundation reauthorization bill and the importance of the MREFC account to the research community and to the health of the US research enterprise, the NSF leadership and the NSB must give careful attention to the implementation of reforms in the MREFC account.

The NSF Authorization Act (HR 4664), which authorizes appropriations for NSF for fiscal years 2003-2007, states the following in Section 14(e):

> (1) STUDY—Not later than 3 months after the date of the enactment of this Act, the Director shall enter into an arrangement with the National Academy of Sciences to perform a study on setting priorities for a diverse array of disciplinary and interdisciplinary Foundation-sponsored large research facility projects.
>
> (2) TRANSMITTAL TO CONGRESS—Not later than 15 months after the date of enactment of this Act, the Director shall transmit to the Committee on Science and the Committee on Appropriations of the House of Representatives and to the Committee on Commerce, Science, and Transportation, the Committee on Health, Education, Labor, and Pensions, and the Committee on Appropriations of the Senate, the study conducted by the National Academy of Sciences together with the Foundation's reaction to the study authorized under paragraph (1).

The committee suggests a six-step process (described in the next section) to implement the recommendations that we hope NSF and NSB will adopt. The committee believes that NSF and NSB reactions to the present report would best be illustrated by a memorandum that would describe the degree to which the report's recommendations and proposed implementation plan are adopted and discuss the extent to which there is agreement or disagreement with the report's recommendations. In addition, NSF and the NSB should describe what actions they plan to undertake to set priorities among NSF's large research facility projects. This report should be provided not only to Congress but to the entire research community for public comment.

Implementing the Recommendations

D escribed below are six steps that the committee believes are necessary to implement its recommendations. Before NSF begins the roadmapping process, it should publish its plans for developing the roadmap. Figure 4 shows the steps in sequence from the development of large research facility project ideas to construction and operation.

STEP 1: Involve the broad research community in identifying, evaluating, and ranking ideas for large facility projects.

To enable the consideration of a wide array of potential large facility projects, NSF should fund on a regular basis, in every field of research that it supports, an assessment of the major research opportunities that require the use of large facilities. Where multiple projects are identified, ranking by the first category of criteria should be used. In some fields of research, particularly fields that overlap disciplines or fall between disciplines, NSF staff should be more active in fostering the planning and assessment of the need for large facility projects.

Some research fields—such as physics, astronomy, and oceans research—already work together to generate ideas for large facility projects. For example, the astronomy community, in an effort funded by NSF and NASA, conducts a decadal study to rank proposed projects. NSF should fund appropriate efforts by other research fields and interdisciplinary areas.

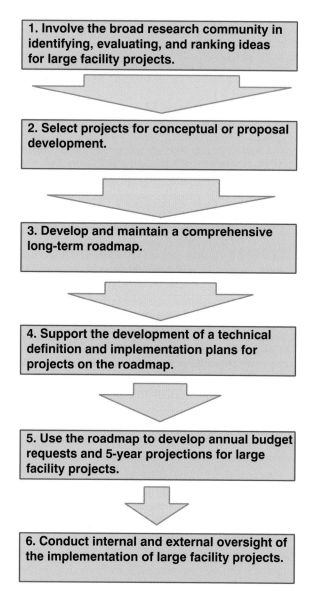

FIGURE 4 Flowchart of enhanced selection, development, and approval process for large facility projects to be supported by NSF. To implement the committee's recommendations, NSF should improve the process whereby large facility projects are identified, developed, and moved forward. The arrows between steps represent the ideas, proposals, or plans for large facility projects progressing from one step to the next; the decreasing width of the arrows represents the successive selection and priority setting among large facility projects.

Broad community involvement at an early stage, including discussion with potential international and interagency partners, is essential to conceiving, identifying, and evaluating ideas for large facility projects.

The NSB, the NSF director, and the heads of the directorates may also generate ideas for large research facilities. But NSF should conduct broad-based community discussions to gauge community need and support for these projects, and the projects should go through the same approval process as community-generated projects.

STEP 2: Select projects for conceptual or proposal development.

Once ideas for large research facilities have been generated and evaluated according to the first group of priority-setting criteria, the research community, via NSF directorate advisory committees, should select projects to move to the next stage of developing conceptual plans or proposals by ranking projects with the second group of priority-setting criteria. Consultation with potential international partners is also useful at this stage.

In deciding which projects to support for initial planning, NSF should make full use of its directorate advisory committees. The members of the science and engineering communities who serve on those committees typically have a broad array of experience and are well positioned to compare projects in a discipline and across related disciplines.[1] In addition, NSF should fund efforts by research communities to set priorities among projects within particular fields and to discuss the need for large facilities versus other kinds of research support. Each conceptual plan or proposal that emerges from this step should include a projected timeline and budget estimate for implementation and operation. This development of project conceptual scope may require limited but specific funding.

STEP 3: Develop and maintain a comprehensive long-term roadmap.

Drawing on the conceptual plans and more detailed proposals for large facility projects across all the research fields that it supports, the NSB should oversee a process using the third group of priority-setting criteria whereby NSF produces a detailed roadmap for ranking large-research-facility projects that it proposes for construction over the next 20 years, as described in Recommendation 1. The roadmap should be reviewed, revised, and updated every 3-5 years and should address specifically how the large facilities it includes will contribute to NSF's

---

[1]For example, the Mathematics and Physical Sciences Directorate includes mathematics, chemistry, astronomy, physics, and materials research.

core mission—"to promote the progress of science; to advance the national health, prosperity, and welfare; and to secure the national defense."[2] The roadmap should also be closely coordinated with the long-range research plans of other agencies and countries.

To be a realistic assessment of future plans, the roadmap must include a projected budgetary envelope for large facility projects. Of course, future budgets cannot be forecasted with certainty, particularly for the long term. But the NSF Reauthorization Act of 2002 established broad outlines for the future growth of NSF's budget, and conservative estimates of continued growth consistent with the policy expressed in the act could be used to establish realistic budget envelopes. By demonstrating the potential of large facility projects to extend knowledge and benefit society, a large facilities roadmap would provide continuing incentives to increase funding of the MREFC account and of the NSF budget in general. Figure 2 represents one possible way to graphically display one of the roadmap outcomes—a proposal for large facility projects phased over the next 20 years within a projected budgetary envelope.

The NSB and NSF need to develop mechanisms to involve the research community broadly in the review and revision of the roadmap. Its construction and iteration should reflect extensive input and feedback from the research community (see Steps 1 and 2). In particular, proponents of individual large facility projects should have opportunities to present their views directly to the NSB and to NSF. The decision process for priority setting in the roadmap is outlined in Recommendation 1. The results and rationale of the roadmapping process should be openly and widely disseminated so that the research community can know what decisions were made and the reasons behind them. A critical companion to the roadmap should be a document describing the process, the stakeholders, and the justification of the roadmap's contents. This transparency in the roadmapping process will help to increase support in the research community for the results of priority setting.

The roadmap will undoubtedly change in response to changing scientific needs during its periodic revisions. Given the roadmap's long-term nature, it will also have to be adjusted to accommodate the realities of short-term budgets. The inclusion of a project on the 20-year roadmap would not necessarily guarantee funding in the president's annual budget requests, in that priorities can change; but it should guide NSF in funding planning and design work to get proposals ready for funding. The readiness of the projects included will depend on whether they are near-term or longer term projects.

---

[2]National Science Foundation Act of 1950.

Figure 5 shows the roles and responsibilities of NSF, the NSB, and the scientific community in the large facility process.

Science and Engineering Research Community

- *Assesses and ranks major research opportunities requiring large facilities*
- *Sets priorities among projects within related fields via NSF directorate advisory committees*
- *Proponents of individual projects present ideas directly to NSB*
- *Comments on draft roadmap*
- *Serves on review committees to oversee project implementation and operation with project management, construction, and operations experts*

National Science Foundation Director and Staff

- *Funds and facilitates science and engineering research community assessments and ranking of major research opportunities requiring large facility projects*
- *Uses guidance from NSF directorate advisory committees to set priorities among projects within related fields*
- *Produces a detailed roadmap for NSB approval*
- *Supports development of technical definition and detailed implementation plans for projects on roadmap*
- *Works with NSB to prepare annual MREFC budget request and rationale for included projects*
- *Deputy for large facility projects oversees implementation and operation of projects*
- *Works with OSTP to coordinate interagency and international collaboration*

National Science Board

- *Oversees and approves roadmap produced by NSF*
- *Reviews and revises roadmap*
- *Works with NSF director and staff to prepare annual MREFC budget request and rationale for included projects*
- *Reviews annual project summaries of each facility's construction and operation*
- *Reviews and approves any proposed major changes in project technical definition*

FIGURE 5 Roles and responsibilities in the large research facility process. The science and engineering community, NSF, and the NSB have different and complementary roles and responsibilities in the process of proposing, constructing, and operating large research facilities. This figure describes their involvement in the process and demonstrates the relationships between them.

STEP 4: Support the development of a technical definition and implementation plans for projects on the roadmap.

Complex projects evolve in phases, starting with a conceptual, creative phase that shifts to engineering and procurement followed by construction and testing operations. Projects on the roadmap will inevitably be at different stages of this process. Guided by the roadmap, NSF should provide specific support for projects to develop a technical definition and implementation plans. The role of the deputy for large facility projects is to manage and oversee this process to ensure successful construction and operation.

As planning proceeds, projects should move from the preconceptual phase to a clear technical definition of what is to be built and what is to be achieved. The greater the analysis in this stage of the process (as opposed to the engineering and construction phase), the more likely that a project will stay within its budget and schedule. The conceptual phase can be only generally budgeted but includes design studies of options, locations, and layouts. Implementation plans should then be developed that describe how the project will be constructed and managed and within what constraints it will perform. Also included should be a description of contracting, scheduling, budgeting, and monitoring processes and provisions for effective cost and schedule control. If key decisions are made as early as possible, project decisions that are disruptive of cost and schedule can be avoided.

Projects in the long-range portion of the roadmap might be limited to a set of scientific or engineering objectives, a rough technical plan, a projected timeline, and a preliminary cost assessment for implementation and operations. But for projects being considered by the NSB for inclusion in NSF's budget, much more detailed plans are necessary, and this planning generally will need to be supported by NSF. For a project being considered as a proposed new start over the following 5 years, detailed implementation plans should be prepared before the NSB approves moving forward. The implementation plans should form the basis of a final evaluation by NSF and the NSB of the quality, feasibility, likely impacts, and importance of the proposed project.

Implementation plans should clearly delineate construction activities funded through the MREFC account and facility operations funded through the Research and Related Activities (R&RA) account, taking special note of the importance of facility testing and commissioning. Provisions for international or interagency coordination also should be clearly laid out in the implementation plans.

Successful project performance requires active leadership of scientists skilled in the elements of the particular sciences that the project embodies. Likewise, large facility research projects require a degree of professionalism and rigorous management experience that researchers cannot single-handedly provide. As a large facility project develops (as described in Step 2), the science and engineering research community's interest in changing the project's scope to incorporate the latest breakthroughs must be tempered by the project management's interest in freezing the design so that an orderly progression of work can be scheduled and budgeted.

STEP 5: Use the roadmap to develop annual budget requests and 5-year projections for large facility projects.

Drawing on the implementation plans and the roadmap, NSF and the NSB should prepare annual budget requests as outlined in Recommendation 2. The budget requests should reflect the ranking embodied in the roadmap and supported by the research community. A budget submission and eventual budget request should encompass the large facility projects to be undertaken over the next 5 years and include a ranked list of the projected new starts and detailed explanations for any deviations from the roadmap.

Fiscal and policy realities may need to be considered in preparation of the budget request and necessitate additional priority setting among the projects recommended by the large facility roadmap. Commitments to projects already under way should be honored first to maintain optimal construction schedules. The final decision-making responsibility rests with the director of NSF and the NSB, who must apply the criteria to assess relative importance as described in Figure 3. Because of the immediacy of their decision, they must choose between projects equally ready for implementation and weigh the costs and benefits of each to render final judgments.

When the administration's budget request is sent to Congress, NSF and the NSB should release documents that explicitly provide the rationale for priority setting among the projects included in the budget. The negotiations between OMB and NSF to determine the administration's budget request will remain confidential, but the resulting mix of large facility projects included in each year's budget should be thoroughly described and explained to create confidence that the roadmap is being implemented to the greatest extent possible. With the information accompanying the budget, Congress has the opportunity to add to the budget if that is desired and to respond better to those who advocate that Congress ignore that information.

STEP 6: Conduct internal and external oversight of the implementation of large facility projects.

NSF makes awards to other organizations to construct, operate, and manage large research facilities. Nevertheless, NSF has the final responsibility for the development, management, and performance of the large facilities it supports. The importance of carefully defining the technical scope and the implementation plan for a project cannot be overemphasized. The time and effort required to achieve a manageable level of understanding or control over the project must not be underestimated. If development is phased from idea generation to conceptual design and engineering planning, a project's final implementation plan should provide a credible path to construction and operation.

To assist with effective management and to assess project performance, large facility projects should be visited and reviewed periodically by panels of outside persons who are experts in the technical subjects that the project entails and experienced in project implementation and management. The review panels should supplement the normal internal project-review activities and can be consulted in overseeing any major changes in implementation plans to ensure that change is managed effectively.

Review panels should be composed of objective experts in each subject who will exercise prudence and provide an independent review and critical evaluation of the project; they should provide a balance of scientific and project-management expertise. The deputy for large facility projects should oversee all project reviews and present a summary to the NSB annually.

The deputy should also monitor the transition from construction to operations to ensure that funds from the proper account are used. This will help to respond to concerns from the NSF inspector general regarding the appropriate use of MREFC and R&RA funds. A project's implementation plan will specify this information and the conditions and schedule for operations, but oversight from the deputy will be essential for a successful transition.

NSF's deputy for large facility projects needs adequate staff and institutional authority to assure the NSF leadership and the NSB that proper project management is in place for each project and that work is progressing on schedule and within budget. Each project will have dedicated leadership, but the deputy for large facility projects has principal responsibility to support the undertakings and for oversight and management. In particular, NSF is encouraged to review the model of large facility project management and oversight that DOE's Office of Science uses through its dedicated Construction Management and Support Division. That division, although serving a larger community, has been successful in balancing the uncertainties of predicting the challenges of building unique experimental facilities and the need for responsible project planning, management, and review.[3]

---

[3]The recommendation of specific best practices for project management is not within the scope of this report, but we encourage NSF to take advantage of the large body of relevant literature available. For instance, the University of California President's Council of the National Laboratories established a Project Management Panel that convened a symposium in November 2002 about "Project Management Excellence"; and in September 2001, NSF hosted a "Large Facility Projects Best Practices Workshop." Those activities were invaluable and provided many resources.

Conclusion

arge research facility projects will continue to constitute a vital component of NSF's science and technology portfolio by enabling researchers to examine previously inaccessible phenomena and answer previously intractable questions. NSF has strengthened the priority-setting process for large research facilities in recent years, partly in response to reports from Congress and other organizations.

NSF now has an opportunity to strengthen the program further by incorporating the preparation of a roadmap into its planning process and by involving the research community more fully in the generation and ranking of ideas for large research facilities.

Making choices among competing proposals from different scientific fields will never be easy, but the recommendations and detailed steps described here can help NSF to excel in this critical part of its mission.

Bibliography

Committee on Science, Engineering, and Public Policy. 2001. Implementing the Government Performance and Results Act for Research: A Status Report. Washington, D.C.: National Academy Press. Available online at http://www.nap.edu/books/0309075572/html/.

Committee on Science, Engineering, and Public Policy. 2000. Experiments in International Benchmarking of U.S. Research Fields. Washington, D.C.: National Academy Press. Available online at http://www.nap.edu/books/0309068983/html/.

Committee on Science, Engineering, and Public Policy. 1999. Evaluating Federal Research Programs: Research and the Government Performance and Results Act." Washington, D.C.: National Academy Press. Available online at http://books.nap.edu/books/0309064309/html/index.html.

Committee on Science, Engineering, and Public Policy. 1996. An Assessment of the National Science Foundation's Science and Technology Centers Program. Washington, D.C.: National Academy Press. Available online at http://www.nap.edu/books/0309053240/html/.

Committee on Science, Engineering, and Public Policy. 1994. Major Award Decisionmaking at the National Science Foundation. Washington, D.C.: National Academy Press. Available online at http://books.nap.edu/books/0309050294/html/index.html.

Committee on Science, Engineering, and Public Policy. 1993. Science, Technology, and the Federal Government: National Goals for a New Era. Washington, D.C.: National Academy Press. Available online at http://www.nap.edu/books/NI000193/html/index.html.

Congressional Research Service. 2002. National Science Foundation: Major Research Equipment and Facility Construction, RS21267. Washington, D.C.

England, Merton. 2003. "Big Science, Big Trouble," Chapter 3, draft manuscript of second volume of his history of NSF.

England, Merton. 1982. A Patron for Pure Science: The National Science Foundation's Formative Years, 1945-57, NSF 82-24, Washington, D.C.: National Science Foundation.

National Science Board. 2001. Guidelines for Setting Priority for Major Research Facilities. NSB-01-202. Available online at http://www.nsf.gov/nsb/meetings/2002/nsb02187/nsb02191.htm.

National Science Board. 1974. Criteria for the Selection of Research Projects by the National Science Foundation. NSB-74-300, as approved by the National Science Board at its 167th meeting, October 17-18. Available online at http://www.nsf.gov/nsf/nsfpubs/guide/criteria.htm.

National Science Board. 1967. Criteria for the Support of Research by the National Science Foundation. NSB-67-133, approved by National Science Board at its 112th meeting, May 18-19. Available online at http://www.nsf.gov/pubs/1967/annualreports/ar_1967_appendix_f.pdf.

National Science Board.1965. Big Science: Policies and Procedures Statement. NSB-79-65, adopted by the National Science Board at its 203rd meeting, January 18-19.

National Science Foundation. 2001. Guidelines for Planning and Managing the Major Research Equipment Account. Available online at http://www.nsf.gov/home/about/mre01.html.

National Science Foundation. 1990. Changes to NSF's Proposal Review System, Notice No. 109. Arlington, Va.

National Research Council. 2003. Large-Scale Biomedical Science: Exploring Strategies for Future Research. National Cancer Policy Board and the Division on Earth and Life Studies, Washington, D.C.: The National Academies Press. Available online at http://www.nap.edu/books/0309089123/html/.

National Research Council. 2002. Progress in Improving Project Management at the Department of Energy: 2002 Assessment. Board on Infrastructure and Constructed Environment, Washington, D.C.: National Academy Press. Available online at http://www.nap.edu/books/0309089093/html/.

National Research Council. 2002. Progress in Improving Project Management at the Department of Energy, 2002 Interim Assessment. Letter Report. Board on Infrastructure and Constructed Environment, Washington, D.C.: National Academy Press. Available online at http://www.nap.edu/books/NI000408/html/.

National Research Council. 2001. Astronomy and Astrophysics in the New Millennium. Board on Physics and Astronomy, Washington, D.C.: National Academy Press. Available online at http://www.nap.edu/books/0309070317/html/.

National Research Council. 2001. Progress in Improving Project Management at the Department of Energy. Board on Infrastructure and Constructed Environment, Washington, D.C.: National Academy Press. Available online at http://www.nap.edu/books/0309082803/html/.

National Research Council. 2001. US Astronomy and Physics: Managing an Integrated Program," Board on Physics and Astronomy, Washington, D.C.: National Academy Press. Available online at http://www.nap.edu/books/0309076269/html/.

National Research Council. 2000. The Owner's Role in Project Management and Preproject Planning. Board on Infrastructure and Constructed Environment, Washington, D.C.: National Academy Press. Available online at http://www.nap.edu/books/0309084253/html/.

National Research Council. 1999. Cooperative Stewardship: Managing the Nation's Multidisciplinary User Facilities for Research with Synchrotron Radiation, Neutrons, and High Magnetic Fields. Commission on Physical Sciences, Mathematics, and Applications, Washington, D.C.: National Academy Press. Available online at http://www.nap.edu/books/0309068312/html/.

National Research Council. 1999. Improving Project Management in the Department of Energy. Board on Infrastructure and Constructed Environment, Washington, D.C.: National Academy Press. Available online at http://www.nap.edu/books/0309066263/html/.

National Research Council. 1998. A Strategy for Research in Space Biology and Medicine into the Next Century. Space Studies Board, Washington, D.C.: National Academy Press. Available online at http://books.nap.edu/books/0309060478/html/index.html.

National Research Council. 1995. Managing the Space Sciences. Space Studies Board, Washington, D.C.: National Academy Press. Available online at http://www.nap.edu/books/NX006980/html/.

US Congress. 1991. Facilities Planning at the National Science Foundation. 102nd Congress, 1st session. Hearings before the Science Subcommittee. Committee on Science, Space and Technology, U.S. House of Representatives. Washington, D.C.: Government Printing Office.

US Congress. 1991. Planning and Priority-Setting in the National Science Foundation. 102nd Congress, 1st Session. Hearings before the Science Subcommittee. Committee on Science, Space and Technology. US House of Representatives. Washington, D.C.: Government Printing Office.

Appendixes

Appendix A

Committee and Professional Staff Biographical Information

William F. Brinkman [NAS] (Chair) is a senior research physicist in the Physics Department at Princeton University. He retired as vice president, research, from Bell Laboratories, Lucent Technologies, Murray Hill, New Jersey, on September 30, 2001. In that position, his responsibilities included the direction of all research to enable the advancement of the technology underlying Lucent Technologies' products. He received his B.S. and Ph.D. in physics from the University of Missouri in 1960 and 1965, respectively. He joined Bell Laboratories in 1966 after spending a year as a National Science Foundation postdoctoral fellow at Oxford University. In 1972, he became head of the Infrared Physics and Electronics Research Department, and in 1974 director of the Chemical Physics Research Laboratory. He held the position of director of the Physical Research Laboratory from 1981 until he moved to Sandia in 1984. He was vice president of Sandia National Laboratories from 1984 until 1987. At that point, he returned to Bell Laboratories to become executive director of the Physics Research Division. In 1993, he became physical sciences research vice president and in January 2000 vice president, research at Bell Laboratories. He is a member of the American Philosophical Society, the National Academy of Sciences (NAS), and the American Academy of Arts and Sciences. He has served on a number of National Academies Committees, including chairmanship of COSEPUP's study on NSF's Science and Technology Centers and the National Research Council's Physics Survey Committee and the Committee on Solid-State Sciences. He served as a member of COSEPUP and the NAS Council. He is past president of the American Physical Society and is chairman of the Laboratories Operations Board of the US Department of Energy. Dr. Brinkman was the recipient of the 1994 George E. Pake Prize.

David H. Auston [NAS, NAE] is president of the Kavli Foundation and the Kavli Institute in Oxnard, California. His career encompasses experience in industry and higher education. He has been a member of the technical staff and department head at AT&T's Bell Laboratories, professor of electrical engineering and applied physics and dean of the School of Engineering and Applied Science at Columbia University, provost of Rice University, and president of Case Western Reserve University. Dr. Auston has contributed to research in lasers, nonlinear optics, and solid-state materials. He is a member of the National Academy of Sciences and the National Academy of Engineering and is a fellow of the American Academy of Arts and Sciences, the Institute of Electrical and Electronic Engineers, the Optical Society of America, and the American Physical Society. A native of Toronto, Canada, Dr. Auston earned a bachelor's and a master's degree in engineering physics and electrical engineering from the University of Toronto and a Ph.D. in electrical engineering from the University of California, Berkeley.

Persis S. Drell is professor and director of research at Stanford Linear Accelerator Center, Stanford University. She received her B.A. in mathematics and physics from Wellesley College in 1977. She received her Ph.D. in atomic physics from the University of California, Berkeley, in 1983 with a precision measurement of parity violation in atomic thallium. She then switched to high-energy experimental physics and worked as a postdoctoral scientist with Lawrence Berkeley National Laboratory on the Mark II experiment at the Stanford Linear Accelerator Center (SLAC). She joined the faculty of the Physics Department at Cornell University in 1988, and there her research focused primarily on experiments at the Cornell Electron Storage Ring, an electron-positron colliding-beam facility. In 2000, she became head of the Cornell high-energy group; in 2001, she was named deputy director of Cornell's Laboratory of Nuclear Studies. In 2002, Dr. Drell accepted a position as professor and director of research at the SLAC. While at Cornell, Dr. Drell studied charm and bottom quarks in an effort to measure the fundamental characteristics of the weak interaction. While on sabbatical in 1998, supported by a fellowship from the Guggenheim Foundation, she formed a collaboration with several members of the Cornell Astronomy Department to perform a critical study of type Ia supernovae and their utility as cosmologic distance markers. In addition to the Guggenheim fellowship, Dr. Drell has been the recipient of a National Science Foundation Presidential Young Investigator Award, and she is a fellow of the American Physical Society.

Alan Dressler [NAS] is an astronomer and member of the scientific staff at the Observatories of the Carnegie Institution in Pasadena, California. He received a B.A. in physics at the University of California (UC), Berkeley and his Ph.D. in astronomy and astrophysics from UC, Santa Cruz. He is a member of the National Academy of Sciences, the American Astronomical Society (AAS), and the International Astronomical Union. Dr. Dressler is an expert in optical studies of galaxies and clusters of galaxies and in the large-scale structure of the universe. His three main subjects of research are the birth and evolution of galaxies; mapping the dark-matter distribution through the peculiar motions of galaxies, leading to the discovery of the "Great Attractor"; and the properties of massive black holes at the centers of galaxies. Dr. Dressler is also the principal investigator of Inamori Magellan Areal Camera and Spectrograph (IMACS), a giant wide-field spectrograph for the Magellan 6.5-m telescope. His scientific research is widely recognized, and he is active in public outreach. He won the AAS Pierce Prize in 1983. He served on the 1991 National Research Council Astronomy and Astrophysics Survey Committee Panel on Policy Opportunities and the Task Group on Space Astronomy and Astrophysics's Panel on Galaxies and Stellar Systems. He chaired the Optical-IR-Ground based Panel for the 2001 Astronomy and Astrophysics Survey. In 1996, Dr. Dressler chaired the Association of Universities for Research in Astronomy's Hubble Space Telescope and Beyond Committee that called for the building of a next-generation space telescope to replace the Hubble; this project, now under way, is now called the James Webb Space Telescope. The committee's report also led to the creation of the Origins Theme at the National Aeronautics and Space Administration (NASA) Office of Space Science, which led to a program to integrate the results of astronomical research in the formation of galaxies to planets, stars, and life. For his leadership in space astronomy, he received NASA's Public Service Medal.

William L. Friend [NAE] is chairman of the University of California's President's Council on the National Laboratories—Los Alamos, Lawrence Livermore, and Berkeley. He graduated summa cum laude in chemical engineering from Polytechnic University and received an M.S. from the University of Delaware in 1958, also in chemical engineering. He retired as executive vice president and director of the Bechtel Group, Inc. (BGI), in 1998 after 41 years in the international engineering and construction industry. His special interests include process design, systems engineering, environmental impacts, Latin America, and engineering education.

He was elected a member of the National Academy of Engineering (NAE) in 1993 and treasurer and member of the NAE Council in 2001. Mr. Friend joined Bechtel in 1977 in Houston in the petroleum and chemical sector. He moved to San Francisco in 1980 and later became a corporate officer, director, group executive vice president, and ultimately member of the Executive Committee of the parent BGI. Before joining Bechtel, Mr. Friend had a distinguished career in process engineering and management, including 5 years (1972-1977) as president and chief executive officer of J. F. Pritchard of Kansas City, Missouri, and 15 years (1957-1972) with The Lummus Company in worldwide operations. He currently serves as a member of the NAE Committee on the Diversity of the Engineering Workforce, as chair of NAE's Finance and Budget Committee, and as a member of the executive committee of the National Research Council Governing Board. He has served on the Committee on Building a Long-Term Environmental Quality Research and Development Program in the US Department of Energy and on the National Aeronautics and Space Administration panel on the Space Station. He is a director of Bechtel National, Inc., and a member of the Board of Consultants of Riggs National Corporation.

Bruce Hevly is associate professor in the Department of History, University of Washington. After receiving a B.A. with majors in history and physics at Carleton College, Northfield, Minnesota, he earned a Ph.D. in the Department of History of Science at Johns Hopkins University and later spent 2 years as a postdoctoral scholar in the Program in History and Philosophy of Science at Stanford University, with sponsorship from the Stanford Linear Accelerator Center. He was laboratory historian at the US Naval Research Laboratory from 1985 to 1987 and joined the University of Washington in 1989. His subjects of special interest are the history of technology; the history of modern physics (especially terrestrial physics); science-technology relationships; science, technology, and the military; and the use of history in science teaching. He is a member of the History of Science Society, the Society for the History of Technology, the British Society for the History of Science, and the American Geophysical Union.

Wesley Huntress is the director of the Carnegie Institution's Geophysical Laboratory. He was associate administrator for space science at the National Aeronautics and Space Administration Headquarters from 1993 to 1998 and director of the Solar System Exploration Division from 1990 to 1992. Before joining the Senior Executive Service, Dr. Huntress had been detailed from the California Institute of Technology's Jet Propulsion Laboratory (JPL) for 2 years as special assistant to the director of the Earth Science and Applications Division. Dr. Huntress began his career at JPL as a National Research Council resident associate in 1968 before joining

JPL permanently. Dr. Huntress has over 100 peer-reviewed publications in astrochemistry. He is a member of the National Research Council's Division Committee on Engineering and Physical Sciences. His current professional memberships include the American Astronautical Society (past president), the American Astronomical Society Division of Planetary Sciences (past chair), and the Planetary Society (president). He received his Ph.D. in chemical physics from Stanford University.

Sir Chris Llewellyn-Smith is director of the UK Atomic Energy Authority's Culham Division, which is responsible for the UK's thermonuclear-fusion program and operates the Joint European Torus. He served as provost and president of University College London from 1999 to 2002, director general of CERN (the European Laboratory for Particle Physics) from 1994 to 1998, and chairman of Oxford Physics from 1987 to 1992. He is a theoretical physicist and has worked on a wide array of topics related to particle-physics experiments. He has also published and spoken widely on aspects of science policy and international scientific collaboration. He has been a fellow of the Royal Society since 1984, and his scientific contributions and leadership have been recognized by awards and honors in seven countries on three continents. He has served on numerous national and international advisory bodies, including the Prime Minister's Advisory Committee on Science and Technology. After receiving a D.Phil. in theoretical physics at Oxford in 1967, he worked briefly in the Lebedev Institute in Moscow and then at CERN and the Stanford Linear Accelerator Center before returning to Oxford in 1974.

Linda J. (Lee) Magid is a professor of chemistry at the University of Tennessee (UT). Her research focuses on physiochemical investigations of micelles and polyelectrolytes in aqueous solutions; techniques used include light scattering, small-angle neutron scattering, neutron spin-echo spectroscopy and nuclear magnetic-resonance spectroscopy. She has served as vice-president for research and graduate studies at the University of Kentucky and is now acting director of the University of Tennessee–Oak Ridge National Laboratory Joint Institute for Neutron Sciences. She has a B.S. in chemistry from Rice University and a Ph.D. in chemistry from the University of Tennessee. She is a fellow of the American Association for the Advancement of Science. She is a member of the National Research Council Board on Physics and Astronomy and serves as chair of its Solid State Sciences Committee. In addition, she serves on the NRC's Research Council Board on Assessment of National Institute of Standards and Technology (NIST) Programs Subpanel for NIST Center for Neutron Research and on the US National Committee for the International Union of Pure and Applied Chemistry. She also served on the NRC's Committee on Developing a Federal Materials Strategy.

Marc Y.E. Pelaez is a private consultant to defense and commercial companies. He provides advice on program development, program execution, technology development, and commercialization. His experience includes 28 years in the US Navy, from which he retired as a rear admiral, and over 5 years as a corporate vice president. From 1996 through 2001, Mr. Pelaez was a vice president of Newport News Shipbuilding, the largest private shipbuilding company in the United States. Initially serving as vice president of engineering, he was responsible for all aspects of the company's engineering endeavors, including a workforce of over 4,500 engineers and designers. Mr. Pelaez retired from the company in December 2001. From 1993 to 1996, as chief of naval research, Mr. Pelaez was responsible for the Department of the Navy's $1.5 billion annual corporate science and technology investment, as well as intellectual-property management, technology transfer policy, and worldwide technology monitoring with offices in London and Tokyo. Furthermore, he had fiduciary responsibility for the Department of the Navy's $9 billion research and development budget. Other Navy assignments included being director of submarine technology at the Defense Advanced Research Projects Agency (DARPA) and commanding officer of the nuclear-powered attack submarine USS *Sunfish* (SSN 649). He has served on the boards of directors of several companies and several professional service organizations. He is a past cochair of the Virginia Research and Technology Advisory Commission, a statewide statutory commission providing advice and recommendations to the governor. Mr. Pelaez is a 1968 graduate of the US Naval Academy and the recipient of numerous awards and decorations, including the Distinguished Service Medal and the Marine Machinery Association's Jack Flanagan Award.

Robert H. Rutford is Excellence in Education Foundation Chaired Professor in Geosciences and former president of the University of Texas at Dallas. He earned a B.A. and an M.A. in geography and a Ph.D. in geology from the University of Minnesota. He moved to the University of South Dakota as an assistant professor in 1967, was promoted to associate professor, and served as chairman of the Geology Department from 1969 to 1972. In 1972, he went to the University of Nebraska-Lincoln to head the Ross Ice Shelf Project, a multi-institutional and international research project in Antarctica. He was also involved in the formation of the Polar Ice Coring Office at Nebraska, a group that focused on ice drilling in both polar regions. In April 1975, he became the director of the Division of Polar Programs at the National Science Foundation (NSF) and directed

NSF-sponsored research in the Arctic and Antarctic. Dr. Rutford returned to the University of Nebraska-Lincoln in 1977 as vice chancellor for research and graduate studies and professor of geology, positions he held until becoming president of the University of Texas at Dallas in May 1982. He served there through August of 1994, when he returned to the faculty and was named by the Board of Regents to the chaired professorship he holds today. Dr. Rutford served as president of the Scientific Committee on Antarctic Research from 1998 to 2002. He is a fellow of the Geological Society of America and the Texas Academy of Science. He is a member of the Arctic Institute of North America, the Nebraska Academy of Science, Sigma Xi, and the American Polar Society.

Joseph H. Taylor [NAS] is James S. McDonnell Distinguished Professor of Physics at Princeton University. He has been a professor of physics since 1980 and dean of the faculty since 1997. He taught at the University of Massachusetts, Amherst, from 1969 to 1980. In 1974, he and a graduate student, Russell A. Hulse, discovered the first binary pulsar, using the radiotelescope at the Arecibo Observatory in Puerto Rico. For that discovery and its contribution to the understanding of gravitation, they were awarded the Nobel Prize in Physics in 1993. In 1978, Dr. Taylor helped to found the Five College Radio Astronomy Observatory. His research on pulsars confirmed Einstein's theory of gravitational waves and thus added to the understanding of the laws governing the universe and giving observational proof of Einstein's theory of relativity. Dr. Taylor has received the Dannie Heineman Prize of the American Astronomical Society and American Institute of Physics, a MacArthur Fellowship, and the Wolf Prize in physics. He is a member of the National Academy of Sciences and the American Philosophical Society and is a fellow of the American Academy of Arts and Sciences and the American Physical Society. He was cochair of the National Research Council's Decade Survey of Astronomy and Astrophysics from 1999 to 2002. He earned his B.A. with honors in physics from Haverford College and his Ph.D. in astronomy from Harvard University. Dr. Taylor's group continues to explore problems in astrophysics and gravitational physics by means of radio-wavelength studies of pulsars. Among recent highlights are the discovery of many new pulsars, including millisecond and binary pulsars.

Michael L. Telson is the director of national laboratory affairs for the University of California in its Washington Office of Federal Governmental Relations. He is responsible for managing the federal regulatory and legislative issues involving the three national laboratories managed by the University of California for the US Department of Energy (DOE): the Lawrence Berkeley, Lawrence Livermore, and Los Alamos laboratories.

He previously served as chief financial officer (CFO) of DOE from October of 1997 (after confirmation by the Senate) through May of 2001. He managed the relationship between the department and the Office of Management and Budget, four congressional appropriations subcommittees, DOE's inspector general, and the General Accounting Office. He reported directly to Secretaries Pena, Richardson, and Abraham, advising them on all financial matters, including the preparation and execution of DOE's nearly $20 billion annual budget, and reprogramming requests in all DOE's business lines, including national security, science, energy, and environmental quality. As CFO, he directed a staff of more than 200, also covering other activities, including project-management oversight, strategic planning and the Government Performance and Results Act, privatization (including the sale of the Elk Hills Naval Petroleum Reserve, the initial public offering of stock in the US Enrichment Corporation, and several environmental management-privatization projects), payroll, and financial-statement issues. Before serving in DOE, he was a senior analyst on the staff of the US House of Representatives Committee on the Budget. He was responsible for reviewing energy, science, and space issues in the federal budget, including the programs of DOE, the National Science Foundation, and the National Aeronautics and Space Administration; governmentwide R&D policy; and some user-fee programs (including Federal Communications Commission spectrum-auction issues). He is a member of Sigma Xi, Tau Beta Pi, and Eta Kappa Nu. He is an American Association for the Advancement of Science fellow and received the Meritorious Service, Superior Performance, and Gold Medal awards for excellence while at DOE. In 2002, he was named a senior fellow of the US Association for Energy Economics. He holds B.S., M.S., and Ph.D. degrees in electrical engineering from the Massachusetts Institute of Technology and an M.S. in management from the Sloan School of Management.

G. David Tilman [NAS] is Regents Professor, holds the McKnight University Presidential Chair in the Department of Ecology, Evolution, and Behavior Ecology, and is the director of Cedar Creek Natural History Area at the University of Minnesota. He was elected to the National Academy of Sciences for his discoveries of how a varied assemblage of species influences the functioning of ecosystems. He has written two books, edited three books, and published more than 160 scientific papers. He is an experimental and theoretical ecologist interested in biological diversity, in the controls of ecosystem stability and productivity, and in the long-term societal implications of human impacts on global ecosystems. For the last 20 years, he has headed the Cedar Creek Long-Term Ecological Research (LTER) Project, one of several National Science Foundation-funded LTER projects nationwide. He is a Guggenheim fellow, a fellow of the American Association for the Advancement of Science, a fellow of the American Academy of Arts and Sciences, a Pew scholar in conservation biology, and a recipient of the Ecological Society of America's Cooper Award and the Robert H. MacArthur award. In 2001, he was designated the most highly cited environmental scientist for the decade 1990-2000 and the decade 1992-2002 by the Institute for Scientific Information. In 1996, he founded *Issues in Ecology* to foster communication among ecologists, the public, and government decision-makers. He has served on the Biodiversity and Ecosystems Panel of the President's Committee of Advisors on Science and Technology (from 1997 to 1998), as a science adviser for Public Radio International's "The World" (from 1997 to 1998), and on the editorial boards of *Science, Proceedings of the National Academy of Sciences, Ecology, Ecological Monographs, The American Naturalist, Acta Oecologia* (Paris), *International Journal of Plant Sciences*, and *Limnology and Oceanography*. He is a member of the National Research Council's Board on Environmental Studies and Toxicology.

PROFESSIONAL STAFF

Deborah D. Stine (Study Director) is associate director of the Committee on Science, Engineering, and Public Policy (COSEPUP) and director of the Office of Special Projects. She has worked on various projects in the National Academies since 1989. She received a National Research Council group award for her first study for COSEPUP, on policy implications of greenhouse warming, a Commission on Life Sciences staff citation for her work in risk assessment and management, and two awards from the Policy and Global Affairs Division for her dissemination efforts for National Academies reports. Other studies have addressed human reproductive cloning, science and technology centers, international benchmarking of US research fields, graduate and postdoctoral education, responsible conduct of research, careers in science and engineering, and many environmental topics. She holds a bachelor's degree in mechanical and environmental engineering from the University of California, Irvine; a master's degree in business administration; and a Ph.D. in public administration, specializing in policy analysis, from the American University. Before coming to the National Academies, she was a mathematician for the US Air Force, an air-pollution engineer for the state of Texas, and an air-issues manager for the Chemical Manufacturers Association.

Timothy I. Meyer is a program officer with the National Research Council's Board on Physics and Astronomy. Dr. Meyer joined the NRC's staff in 2002 after earning his Ph.D. in experimental particle physics from Stanford University. His thesis concerned the time evolution of the B meson in the

BaBar experiment at the Stanford Linear Accelerator Center. His work also focused on radiation monitoring and protection of silicon-based particle detectors. During his time at Stanford, Dr. Meyer received both the Paul Kirkpatrick and Centennial Teaching awards for his work as an instructor of undergraduates. He is a member of the American Physical Society, the American Association for the Advancement of Science, Phi Beta Kappa, and the Union of Concerned Scientists.

Donald C. Shapero is director of the National Research Council's Board on Physics and Astronomy. He received a B.S. from the Massachusetts Institute of Technology (MIT) in 1964 and a Ph.D. from MIT in 1970. His thesis addressed the asymptotic behavior of relativistic quantum field theories. After receiving the Ph.D., he became a Thomas J. Watson postdoctoral fellow at IBM. He subsequently became an assistant professor at American University, later moving to Catholic University and then joining the staff of the NRC in 1975. He took a leave of absence from the NRC in 1978 to serve as the first executive director of the Energy Research Advisory Board at the Department of Energy. He returned in 1979 to serve as special assistant to the president of the National Academy of Sciences. In 1982, he started the Board on Physics and Astronomy (BPA). As BPA director, he has played a key role in many NRC studies, including the two most recent surveys of physics and the two most recent surveys of astronomy and astrophysics. He is a member of the American Physical Society, the American Astronomical Society, and the International Astronomical Union. He has published research articles in refereed journals in high-energy physics, condensed-matter physics, and environmental science.

Richard E. Bissell joined the National Academies in June 1998 as executive director of the Policy Division, now the Policy and Global Affairs Division, and concurrently as director of the Committee on Science, Engineering, and Public Policy. He most recently headed the organizing secretariat of the World Commission on Dams, a joint initiative of the World Bank and the World Conservation Union. During 1994-1997, he was a founding member and chair of the Inspection Panel, an independent accountability mechanism established by the executive directors of the World Bank to ensure compliance with Bank policies by its management. During the years 1986-1993, he was assistant administrator at the US Agency for International Development, first as head of the Bureau of Policy and Program Coordination and later as head of the Bureau of Research and Development. His B.A. is from Stanford, and his Ph.D. from the Fletcher School of Law and Diplomacy at Tufts University.

Appendix B

Charge to the Committee and Origins of the Study

STATEMENT OF TASK

In response to a Congressional request, the committee conducting this study will examine how the National Science Foundation sets priorities among multiple competing proposals for construction and operation of large-scale research facility projects for a diverse array of disciplines, and will make recommendations regarding how to make the priority-setting process as effective as possible, taking into account NSF's significant role in funding academic research in science and engineering in the United States.

Specifically, the committee will address the following tasks:

- Review NSF's current prioritization process as well as processes and procedures used by other relevant organizations.
- Develop the criteria that should be considered in developing priorities among competing large research facility proposals.
- Provide recommendations for optimizing and strengthening the process used by the NSF to set priorities among large research facility project proposals and to manage their incorporations into the President's budget.
- Provide recommendations for improving the construction and operation of NSF-funded large research facility projects.
- Provide recommendations regarding the role of the current and future availability of international and interagency research facility projects in the decision-making process for NSF funding of large research facility projects.

LETTER FROM US CONGRESS

Included on the next two pages is a reproduction of a letter from several US senators to Bruce Alberts, president of the National Academy of Sciences, requesting this study.

𝔘nited 𝔖tates 𝔖enate

WASHINGTON, DC 20510

June 12, 2002

Dr. Bruce Alberts
President
National Academy of Sciences
2101 Constitution Ave, NW
Washington, D.C. 20418

Dear Dr. Alberts:

By this letter, we request the National Academy of Sciences (NAS) to develop a set of criteria that can be used to rank and prioritize large research facility projects sponsored by the National Science Foundation (NSF) – particularly those funded through the Major Research Equipment and Facilities Construction account. Despite several efforts, questions remain as to whether NSF has a satisfactory process for prioritizing multiple competing large-scale research facility proposals. As a result, funding requests by the Foundation for large facility projects appear to be ad hoc and subjective.

In recent years, with congressional support, NSF has increased its investments in large infrastructure projects such as accelerators, telescopes, research vessels, supercomputers, digital databases, and earthquake simulators. NSF spends approximately $1 billion per year for such cutting-edge projects, some of which individually cost hundreds of millions of dollars. Many of these projects are large in scale, require complex instrumentation, and involve partnerships with other Federal agencies, international science organizations, and foreign governments.

We request the NAS to review the current prioritization process and report to us on how it can be improved. Specifically, we ask that you provide us with specific criteria that will lead to a prioritized ranking of competing large research facility proposals that address both scientific merit and management criteria. We ask that you consider project management capability as a criterion because NSF heavily relies on the management capabilities of its awardees to construct and operate its large facility projects.

We also believe NSF should play a stronger role in the management, oversight, and accountability of the projects that it ultimately supports. The NSF Inspector General has recently found significant deficiencies in the Foundation's management and oversight of its large facility projects resulting in significant cost overruns not contemplated in their original budgets. We request the Academy to provide us with recommendations to help the Foundation address this issue.

Lastly, we are interested in the Academy's views about the availability of large research facilities in other countries. For some types of scientific research, existing overseas facilities may be adequate and cost-effective in meeting U.S. research needs through international partnerships. We ask that you consider this issue as a possible criterion for a prioritized ranking system.

Thank you for your consideration of this request. Please contact Cheh Kim of the VA-HUD Subcommittee staff at 202-224-7858 if you have any questions.

Sincerely,

Barbara A. Mikulski
Chair
Subcommittee on VA, HUD,
   and Independent Agencies

Christopher S. Bond
Ranking Member
Subcommittee on VA, HUD,
   and Independent Agencies

Ernest F. Hollings
Chair
Committee on Commerce,
   Science, and Technology

John McCain
Ranking Member
Committee on Commerce,
   Science, and Technology

Edward M. Kennedy
Chair
Committee on Health, Education,
   Labor, and Pensions

Judd Gregg
Ranking Member
Committee on Health, Education,
   Labor, and Pensions

LANGUAGE FROM HR 4664, NSF AUTHORIZATION ACT OF 2002

The NSF budget authorization bill of 2002 carried with it specific language that instructed the National Academy of Sciences to undertake this study. The final version of the authorization bill was signed into law on December 19, 2002 as Public Law 107-368. Section 14 of the bill discusses the Major Research Equipment and Facilities Construction Plan; this section is reproduced on the following pages for reference.

One Hundred Seventh Congress
of the
United States of America

AT THE SECOND SESSION

Begun and held at the City of Washington on Wednesday,
the twenty-third day of January, two thousand and two

An Act

To authorize appropriations for fiscal years 2003, 2004, 2005, 2006, and 2007
for the National Science Foundation, and for other purposes.

Be it enacted by the Senate and House of Representatives of
the United States of America in Congress assembled,

SECTION 1. SHORT TITLE.

This Act may be cited as the "National Science Foundation
Authorization Act of 2002".

SEC. 2. FINDINGS.

Congress finds the following:

(1) The National Science Foundation has made major contributions for more than 50 years to strengthen and sustain
the Nation's academic research enterprise that is the envy
of the world.

(2) The economic strength and national security of the
United States and the quality of life of all Americans are
grounded in the Nation's scientific and technological capabilities.

(3) The National Science Foundation carries out important
functions in supporting basic research in all science and
engineering disciplines and in supporting science, mathematics,
engineering, and technology education at all levels.

(4) The research and education activities of the National
Science Foundation promote the discovery, integration, dissemination, and application of new knowledge in service to society
and prepare future generations of scientists, mathematicians,
and engineers who will be necessary to ensure America's leadership in the global marketplace.

(5) The National Science Foundation must be provided
with sufficient resources to enable it to carry out its responsibilities to develop intellectual capital, strengthen the scientific
infrastructure, integrate research and education, enhance the
delivery of mathematics and science education in the United
States, and improve the technological literacy of all people
in the United States.

(6) The emerging global economic, scientific, and technical
environment challenges long-standing assumptions about
domestic and international policy, requiring the National
Science Foundation to play a more proactive role in sustaining
the competitive advantage of the United States through superior research capabilities.

H. R. 4664—23

the Foundation's reaction to the assessment authorized under paragraph (1).

SEC. 14. MAJOR RESEARCH EQUIPMENT AND FACILITIES CONSTRUCTION PLAN.

(a) PRIORITIZATION OF PROPOSED MAJOR RESEARCH EQUIPMENT AND FACILITIES CONSTRUCTION.—

(1) DEVELOPMENT OF PRIORITIES.—(A) The Director shall—

(i) develop a list indicating by number the relative priority for funding under the major research equipment and facilities construction account that the Director assigns to each project the Board has approved for inclusion in a future budget request; and

(ii) submit the list described in clause (i) to the Board for approval.

(B) The Director shall update the list prepared under subparagraph (A) each time the Board approves a new project that would receive funding under the major research equipment and facilities construction account, as necessary to prepare reports under paragraph (2), and, from time to time, submit any updated list to the Board for approval.

(2) ANNUAL REPORT.—Not later than 90 days after the date of enactment of this Act, and not later than each June 15 thereafter, the Director shall transmit to the Committee on Science of the House of Representatives, the Committee on Commerce, Science, and Transportation of the Senate, and the Committee on Health, Education, Labor, and Pensions of the Senate a report containing—

(A) the most recent Board-approved priority list developed under paragraph (1)(A);

(B) a description of the criteria used to develop such list; and

(C) a description of the major factors for each project that determined the ranking of such project on the list, based on the application of the criteria described pursuant to subparagraph (B).

(3) CRITERIA.—The criteria described pursuant to paragraph (2)(B) shall include, at a minimum—

(A) scientific merit;

(B) broad societal need and probable impact;

(C) consideration of the results of formal prioritization efforts by the scientific community;

(D) readiness of plans for construction and operation;

(E) the applicant's management and administrative capacity of large research facilities;

(F) international and interagency commitments; and

(G) the order in which projects were approved by the Board for inclusion in a future budget request.

(b) FACILITIES PLAN.—

(1) IN GENERAL.—Section 201(a)(1) of the National Science Foundation Authorization Act of 1998 (42 U.S.C. 1862l(a)(1)) is amended to read as follows:

"(1) IN GENERAL.—The Director shall prepare, and include as part of the Foundation's annual budget request to Congress, a plan for the proposed construction of, and repair and upgrades to, national research facilities, including full life-cycle cost information.".

H. R. 4664—24

(2) CONTENTS OF PLAN.—Section 201(a)(2) of the National Science Foundation Authorization Act of 1998 (42 U.S.C. 1862l(a)(2)) is amended—

(A) in subparagraph (A), by striking "(1);" and inserting "(1), including costs for instrumentation development;";

(B) in subparagraph (B), by striking "and" after the semicolon;

(C) in subparagraph (C), by striking "construction." and inserting "construction;"; and

(D) by adding at the end the following:

"(D) for each project funded under the major research equipment and facilities construction account—

"(i) estimates of the total project cost (from planning to commissioning); and

"(ii) the source of funds, including Federal funding identified by appropriations category and non-Federal funding;

"(E) estimates of the full life-cycle cost of each national research facility;

"(F) information on any plans to retire national research facilities; and

"(G) estimates of funding levels for grants supporting research that will be conducted using each national research facility.".

(3) DEFINITION.—Section 2 of the National Science Foundation Authorization Act of 1998 (42 U.S.C. 1862k note) is amended—

(A) by redesignating paragraphs (3) through (5) as paragraphs (4) through (6), respectively; and

(B) by inserting after paragraph (2) the following:

"(3) FULL LIFE-CYCLE COST.—The term 'full life-cycle cost' means all costs of planning, development, procurement, construction, operations and support, and shut-down costs, without regard to funding source and without regard to what entity manages the project or facility involved.".

(c) PROJECT MANAGEMENT.—No national research facility project funded under the major research equipment and facilities construction account shall be managed by an individual whose appointment to the Foundation is temporary.

(d) BOARD APPROVAL OF MAJOR RESEARCH EQUIPMENT AND FACILITIES PROJECTS.—

(1) IN GENERAL.—The Board shall explicitly approve any project to be funded out of the major research equipment and facilities construction account before any funds may be obligated from such account for such project.

(2) REPORT.—Not later than September 15 of each fiscal year, the Board shall report to the Committee on Commerce, Science, and Transportation of the Senate, the Committee on Health, Education, Labor, and Pensions of the Senate, and the Committee on Science of the House of Representatives on the conditions of any delegation of authority under section 4 of the National Science Foundation Act of 1950 (42 U.S.C. 1863) that relates to funds appropriated for any project in the major research equipment and facilities construction account.

(e) NATIONAL ACADEMY OF SCIENCES STUDY ON MAJOR RESEARCH EQUIPMENT AND FACILITIES CONSTRUCTION.—

H. R. 4664—25

(1) STUDY.—Not later than 3 months after the date of enactment of this Act, the Director shall enter into an arrangement with the National Academy of Sciences to perform a study on setting priorities for a diverse array of disciplinary and interdisciplinary Foundation-sponsored large research facility projects.

(2) TRANSMITTAL TO CONGRESS.—Not later than 15 months after the date of the enactment of this Act, the Director shall transmit to the Committee on Science and the Committee on Appropriations of the House of Representatives, and to the Committee on Commerce, Science, and Transportation, the Committee on Health, Education, Labor, and Pensions, and the Committee on Appropriations of the Senate, the study conducted by the National Academy of Sciences together with the Foundation's reaction to the study authorized under paragraph (1).

SEC. 15. ADMINISTRATIVE AMENDMENTS.

(a) BOARD MEETINGS.—

(1) IN GENERAL.—Section 4(e) of the National Science Foundation Act of 1950 (42 U.S.C. 1863(e)) is amended by striking the second and third sentences and inserting "The Board shall adopt procedures governing the conduct of its meetings, including delivery of notice and a definition of a quorum, which in no case shall be less than one-half plus one of the confirmed members of the Board.".

(2) OPEN MEETINGS.—The Board and all of its committees, subcommittees, and task forces (and any other entity consisting of members of the Board and reporting to the Board) shall be subject to section 552b of title 5, United States Code.

(3) COMPLIANCE AUDIT.—The Inspector General of the Foundation shall conduct an annual audit of the compliance by the Board with the requirements described in paragraph (2). The audit shall examine the proposed and actual content of closed meetings and determine whether the closure of the meetings was consistent with section 552b of title 5, United States Code.

(4) REPORT.—Not later than February 15 of each year, the Inspector General of the Foundation shall transmit to the Committee on Science of the House of Representatives, the Committee on Commerce, Science, and Transportation of the Senate, and the Committee on Health, Education, Labor, and Pensions of the Senate the audit required under paragraph (3) along with recommendations for corrective actions that need to be taken to achieve fuller compliance with the requirements described in paragraph (2), and recommendations on how to ensure public access to the Board's deliberations.

(b) CONFIDENTIALITY OF CERTAIN INFORMATION.—Section 14(i) of the National Science Foundation Act of 1950 (42 U.S.C. 1873(i)) is amended to read as follows:

"(i)(1)(A) Information supplied to the Foundation or a contractor of the Foundation in survey forms, questionnaires, or similar instruments for purposes of section 3(a)(5) or (6) by an individual, an industrial or commercial organization, or an educational, academic, or other nonprofit institution when the institution has received a pledge of confidentiality from the Foundation, shall not

LANGUAGE FROM SENATE REPORT 107-222 TO ACCOMPANY
S. 2797, FY 2003 APPROPRIATIONS BILL FOR
VA/HUD/INDEPENDENT AGENCIES

Senate Report 107-222 includes a discussion of NSF's MREFC account.
The relevant portion of the report on Title III of the appropriations bill is
reproduced on the next four pages for reference.

Calendar No. 519

107TH CONGRESS		REPORT
2d Session	SENATE	107–222

DEPARTMENTS OF VETERANS AFFAIRS AND HOUSING AND URBAN DEVELOPMENT, AND INDEPENDENT AGENCIES APPROPRIATIONS BILL, 2003

JULY 25, 2002.—Ordered to be printed

Ms. MIKULSKI, from the Committee on Appropriations, submitted the following

REPORT

[To accompany S. 2797]

The Committee on Appropriations reports the bill (S. 2797) making appropriations for the Departments of Veterans Affairs and Housing and Urban Development, and for sundry independent agencies, boards, commissions, corporations, and offices for the fiscal year ending September 30, 2003, and for other purposes, reports favorably thereon and recommends that the bill do pass.

Amount of new budget (obligational) authority

Amount of bill as reported to Senate	$124,507,956,000
Amount of appropriations to date, 2002	119,907,308,000
Amount of budget estimates, 2003	121,358,580,000
Over estimates for 2003	3,149,376,000
Above appropriations for 2002	4,600,648,000

127

The Committee has also increased the request for U.S. polar research programs by $10,000,000 to support priority research and infrastructure needs.

As a key part of the Administration's climate change research initiative, the Committee recognizes the Nation needs substantially better information on the current and future state of the ocean and its role in environmental change. Adequate predictive capability is a prerequisite to the development of sound policies at the national and regional level, policies ranging from maritime commerce to public health, from fisheries to safety of life and property, from climate change to national security. The Committee urges NSF to move ahead to support an ocean observatories initiative that is tightly integrated with the Administration's interagency climate change science program.

The Committee supports the fiscal year 2003 budget request for the social, behavioral and economic sciences. Within this amount, the Committee provides $10,000,000 for the children's research initiative.

The Committee is providing an additional $50,000,000 to augment the request for the major research instrumentation program. The Committee reiterates its long-standing concern about the infrastructure needs of developing institutions, historically black colleges and universities; and other minority-serving colleges and universities. The Committee directs NSF to use these additional funds to support the merit-based instrumentation and infrastructure needs of these institutions.

The Committee's recommendation includes an additional $10,000,000 for the innovation partnership program. With these funds, NSF is to support competitive, merit-based partnerships, consisting of States, local and regional entities, industry, academic institutions, and other related organizations for innovation-focused local and regional technology development strategies.

MAJOR RESEARCH EQUIPMENT AND FACILITIES CONSTRUCTION

Appropriations, 2002	$138,800,000
Budget estimate, 2003	126,280,000
Committee recommendation	79,280,000

PROGRAM DESCRIPTION

The major research equipment and facilities construction appropriation supports the acquisition, procurement, construction, and commissioning of unique national research platforms, research resources and major research equipment. Projects supported by this appropriation will push the boundaries of technology and will offer significant expansion of opportunities, often in new directions, for the science and engineering community. Preliminary design and development activities, and on-going operations and maintenance costs of the facilities are provided through the research and related activities appropriation account.

COMMITTEE RECOMMENDATION

The Committee recommends $79,280,000 for major research equipment and facilities construction. Support for the terascale computing systems has been provided in the Research and Related

128

Activities Appropriations Account. Within this account, the Committee's recommendation includes funding for the following projects:

$20,000,000 for Earthscope; $30,000,000 for the Atacama Large Millimeter Array telescope; $9,720,000 for the Large Hadron Collider; $13,560,000 for the Network for Earthquake Engineering Simulation; and $6,000,000 for South Pole Station.

The Committee remains concerned about the Foundation's management of large scale construction projects and the priority setting process used to select projects to be funded. The Committee received a report from NSF required by Public Law 107–73 which addressed a number of issues of concern to the Committee. However neither the report nor the budget justifications addressed the way in which criteria are used by the agency and the National Science Board in setting priorities among new and potential new starts. A recent audit by the Inspector General identified a number of issues in both the financial management and project management of previously funded projects. In addition, the National Academy of Sciences has recently been asked by the Committee and NSF's authorizing committees to assist in the development of a process for prioritizing projects to be funded out of this account. Accordingly, the Committee directs NSF to provide $750,000 to support the Academy's work on this matter. These funds should be made available from resources used for Planning and Evaluation.

The Committee also supports provisions under consideration by the authorizing committees to establish a more transparent process for the establishment of priorities with respect to the funding of major research equipment and facilities construction. The Committee believes a more open and understandable process, which includes National Science Board and NSB Committee meetings, are important aspects of such a priority setting process.

In addition, despite repeated concerns expressed by the Congress and the Inspector General, NSF has not addressed adequately the management and funding problems associated with large research facilities funded through the major research equipment and facilities construction account (formerly named the major research equipment or MRE account). The Inspector General's May 1, 2002 report found that the lack of adequate guidance "have allowed NSF to use multiple appropriation accounts to fund the acquisition and construction costs of major research equipment and facilities, and led to inconsistencies in the types of costs funded through the MRE account." This practice has led to the use of funds from the research and related activities account to pay for cost overruns and scope increases of large facility projects without adequate notification and consultation with the Committee. Accordingly, the Committee directs NSF to include in its fiscal year 2003 operating plan to the Committee a report that details approved budgeted and actual expenditure information on each individual large research facility projects approved by the Congress. The report should include information on the amount of funds approved by the Congress from its inception by year, the amount of actual funds spent on the project by year, and a breakdown of the budgeted and actual expenditures by appropriation account. In addition, the Committee notes the findings and recommendations contained in the OIG re-

129

port pertaining to NSF's cost accounting system. As a result, the Committee also directs NSF to address the deficiencies in its cost accounting system to ensure that the system is capable of readily and reliably providing the Foundation and the Committee with information on the actual cost of NSF programs and activities.

The Committee notes that since last year, the Foundation has been recruiting for a Deputy Director for Large Facility Projects. However, NSF has not yet filled this important position. Accordingly, while the Committee has recommended start up funding for the Earthscope project, bill language has been included delaying the obligation of these funds until NSF fills the position of Deputy Director for Large Facility Projects on a permanent basis.

The Committee notes that NSF is proposing to spend $40,000,000 over the next 3 years to develop two National Ecological Observatory Network (NEON) sites. The Committee notes that NSF considers this the first phase of NEON. Information on the full NEON concept, including cost estimates, has yet to be provided to the Committee. In the absence of such information, and without prejudice, the Committee is not prepared to recommend funding for NEON at this time.

The Committee urges NSF to continue moving forward with the IceCube Neutrino Detector Observatory. The technology developed by IceCube's precursor project has proven successful at detecting high-energy atmospheric neutrinos. Continued development is expected to lead to a new era in astronomy in which researchers will have unique opportunities to analyze some of the most distant and significant events in the formulation and evolution of the universe.

EDUCATION AND HUMAN RESOURCES

Appropriations, 2002	$875,000,000
Budget estimate, 2003	908,080,000
Committee recommendation	947,730,000

PROGRAM DESCRIPTION

The education and human resources appropriation supports a comprehensive set of programs across all levels of education in science, technology, engineering and mathematics (STEM). The appropriation supports activities that unite school districts with institutions of higher learning to improve precollege education. Other precollege activities include development of the next generation of precollege STEM education leaders; instructional materials; and the stem instructional workforce. Undergraduate activities support curriculum, laboratory, and instructional improvement; expand the STEM talent pool through scholarships and attracting STEM participants to teaching; augment advanced technological education at 2-year colleges; and develop dissemination tools. Graduate support is directed to research and teaching fellowships and traineeships, and linking precollege systems with higher education to improve the instructional workforce. Programs also seek to broaden the participation of groups underrepresented in the STEM enterprise; build State and regional capacity to compete successfully for research funding; and promote informal science education. Ongoing evaluation efforts and research on learning strengthen the base for these programs. In addition to this appropriation, the Foundation

Appendix C

Histories of Projects Funded by National Science Foundation Major Research Equipment and Facilities Construction Account

T his appendix briefly describes the projects approved for construction funding through the National Science Foundation (NSF) Major Research Equipment and Facilities Construction (MREFC) account. For each project, the committee provides a brief description and a timeline of major developments. Project descriptions and funding information for all funded projects were reviewed by NSF staff.

The following projects are described:

WHAT IS A LARGE FACILITY PROJECT?

In FY 1995, NSF created what is now known as the Major Research Equipment and Facilities Construction account to support the "acquisition, construction, commissioning, and upgrading of major research equipment, facilities, and other such capital assets" that cost more than several tens of millions of dollars.[1] As of September 2003, the account has funded 12 large facility projects, and four new projects are proposed in NSF's FY 2004 budget request to receive funding. Note that in some cases there was or is a gap in funding.

The projects listed below have been, are being, or are proposed to be supported by the MREFC account. They appear with the fiscal year in which construction funding began or is proposed to begin.

Construction Projects Supported in the Past:
- Laser Interferometer Gravitational-Wave Observatory (LIGO)—FY 1992
- Gemini Observatories—FY 1991
- Polar Support Aircraft Upgrades—FY 1999
- South Pole Safety and Environmental Project (SPSE)—FY 1997
- Terascale Computing Projects—FY 2000

Construction Projects Currently Being Supported:
- South Pole Station Modernization (SPSM)—FY 1998
- Large Hadron Collider (LHC)—FY 1999
- Network for Earthquake Engineering Simulation (NEES)—FY 2000
- Atacama Large Millimeter Array/Millimeter Array (ALMA/MMA)—FY 1998
- EarthScope—FY 2003
- IceCube Neutrino Detector—FY 2002

Initiated Projects Currently Experiencing a Gap in MRE Funding:
- High-Performance Instrumented Airborne Platform for Environmental Research (HIAPER)—FY 2000

---

[1]Congressional Research Service, Library of Congress, *National Science Foundation: Major Research Equipment and Facility Construction* (Washington, D.C.: Congressional Research Service, 2002).

New Starts Proposed in NSF's FY 2004 Budget for FY 2004, 2005, or 2006 Support:
- National Ecological Observatory Network (NEON) Phase I—FY 2004
- Rare Symmetry Violating Processes (RSVP)—FY 2006
- Ocean Observatories Initiative (OOI)—FY 2006
- Integrated ocean drilling program (IODP)—FY 2005

ALMA (ATACAMA LARGE MILLIMETER ARRAY)

Description

The Atacama Large Millimeter Array (ALMA) will be a 64-element array of 12-m-diameter radio antennas in the Chilean Andes. The array is designed to study the millimeter- and submillimeter-wavelength portions of the spectrum with "unprecedented imaging capabilities and sensitivity many orders of magnitude greater than anything of its kind today." [1] The principal contributors to the development and construction of ALMA are the National Radio Astronomy Observatory (NRAO) and the European Southern Observatory (ESO), but many other international partners are involved.

See Table C-1 for a timeline of the major developments.

Approval and Funding History

MREFC funding for planning, design, and development began in FY 1998; this stage of the project is referred to as ALMA I. MREFC funding for construction began in FY 2002; the construction phase is referred to as ALMA II.

Managing Institutions

ALMA is an international collaboration. The US side of the project is led by Associated Universities, Inc., and the NRAO. Europe is an equal partner in ALMA with funding and execution carried out through the ESO.

Development Summary

Millimeter Array

In the spring of 1982, it was recognized that a proposal for a 25-m dish for millimeter astronomy initiated by the NRAO in 1975 [2] might never be funded [3, 4]. Robert Wilson called for a meeting at Bell Telephone Laboratories (BTL) in October 1982, intentionally excluding NRAO scien-

TABLE C-1 Timeline of Major Developments

1975	NRAO proposes 25-m dish for millimeter astronomy to be built on Mauna Kea in Hawaii [2].
1981	NRAO astronomers begin initial design work for millimeter-wavelength array [6, 7].
Spring 1982	Astronomy community recognizes that proposed 25-m dish will never be funded [3, 4].
October 1982	First meeting of BTL working group, attended by 18 scientists [3].
December 1982	First meeting of NSF Subcommittee on Millimeter- and Submillimeter-Wavelength Astronomy in Washington, D.C. (Alan Barrett, chair) [3].
February 1983	Joint meeting of BTL working group, Barrett subcommittee, and others to discuss scientific details of new facility [3].
April 1983	Final Barrett subcommittee meeting in Chicago [3]. Barrett subcommittee report is sent to Astronomy Advisory Committee, which endorses recommendation to do design study for millimeter array and passes it on to NSF Division of Astronomical Sciences [3].
1984	Design study for MMA begins [2].
Fall 1985	First MMA science workshop at Green Bank [2, 5].
November 1989	Second MMA science workshop to update scientific goals and array design in preparation for MMA construction proposal [5].
September 1990	Associated Universities, Inc. submits MMA proposal to NSF [5].
May 1991	National Research Council's *The Decade of Discovery in Astronomy and Astrophysics* recommends MMA second among new ground-based initiatives.
October 1991	Two-stage approach for MMA is endorsed by NSF Advisory Committee for the Astronomical Sciences: development phase (detailed designs and prototypes) and construction phase [5, 8].
March 1992	NSF Division of Astronomical Sciences requests 3-year plan for detailed design of MMA [5].
September 1992	MMA detailed design plan is submitted to NSF [5, 8].
November 18, 1994	NSB approves NRAO's project-development plan for MMA [5, 8].
April 1995	NRAO begins site testing in high altitude Atacama Desert in Chile [12].
June 1995	NRAO and Japanese astronomers sign memorandum of understanding to jointly investigate Chilean sites [5].

continued

TABLE C-1 Continued

October 1995	At MMA science workshop, it is concluded that array should have larger baseline and include submillimeter capability [6, 8]; these enhancements would require larger site and higher standards of atmospheric quality than original concept [6].
June 1997	ESO and NRAO sign agreement to explore merging of Large Southern Array (LSA) and MMA; three joint working groups are established to study merger: Science Working Group, Technical Working Group, and Management Working Group [9].
Fall 1997	Congress approves funding for MMA design and development, expected to last 3 years [5].
December 1997	Technical workshop is held to examine possibility of merging MMA and Large Millimeter-Submillimeter Array [5].
April 1998	LSA and MMA feasibility study is completed [9].
May 1998	NRAO report recommends that MMA be built in Atacama Desert [6].
June 1998	Phase 1: Research and development of MMA project begins [5] after NSB authorization.
June 1999	US-European memorandum of understanding is signed, merging two Phase 1 projects into ALMA [1].
2000	National Research Council Astronomy and Astrophysics Survey Committee reaffirms its 1991 endorsement of ALMA.
2002	ALMA receives MREFC funding.
January 24, 2002	NSB Executive Committee authorizes full construction of ALMA [15].
Fall 2002	Prototype antenna testing begins in New Mexico [11].
February 25, 2003	Rita R. Colwell (director, NSF) and Catherine Cesarsky (director general, ESO) sign agreement to jointly construct and operate ALMA [10].

tists, to decide the next step for the millimeter-astronomy community [2, 3]. At the same time, the NSF Advisory Committee for Mathematical and Physical Sciences (MPS/AC) formed a Subcommittee on Millimeter- and Submillimeter-Wavelength Astronomy,[2] chaired by Alan H. Barrett. All five members of the Barrett subcommittee were also members of the BTL

---

[2]Subcommittee members: Alan H. Barrett (Chair), Charles J. Lada, Patrick Palmer, Lewis E. Snyder, and William J. Welch.

working group, and there was a great deal of cooperation between the two groups [3], but the meetings of the Barrett subcommittee were attended by NRAO astronomers. In April 1983, the Barrett subcommittee recommended that a design study for a millimeter array be undertaken, and the NSF MPS/AC passed the recommendation on to NSF [3, 5].

The design work for what would become known as the Millimeter Array (MMA) had in fact started in 1981 at the NRAO [6, 7]. The NSF design study began in 1984, and a gradual community consensus emerged that the NRAO should handle the project [2]. A series of communitywide workshops were held in 1985, 1987, and 1989 [8]. At the first workshop, in the fall of 1985, the scientific goals and design characteristics were discussed. A design concept for the MMA was developed and further refined in the later workshops [2, 5].

In September 1990, Laura Bautz, director of the NSF Division of Astronomical Sciences (AST), received a proposal for the MMA from the NRAO astronomers [2, 5]. The proposal called for an array of 40 8-m antennae with a total collecting area of 2,010 m^2 [2]. In May 1991, the National Research Council report *The Decade of Discovery in Astronomy and Astrophysics* recommended the MMA as a second priority among new ground-based initiatives [13]. In October 1991, the NSF Advisory Committee for Mathematical and Physical Sciences endorsed a plan for the MMA to proceed in two stages: a development phase, in which key equipment would be designed and prototyped, and then a construction phase [5, 8]. A few months later, the NSF AST requested a 3-year plan for a development program, which it received in September 1992 [5]. In November 1994, the NSB approved the project-development plan for the MMA, which demonstrated a scientific need for the facility and embraced a two-stage process to design and build it: a formal three-year design and development phase to be followed by construction, subject to a separate approval by the NSB.

Site Selection

Sites for the MMA were initially considered in Arizona and New Mexico in 1985 when site evaluation and testing began. As a point of reference, similar testing equipment was set up at the Caltech Submillimeter Observatory on Mauna Kea in Hawaii. The advantages of the North American sites were their affordability and location in the United States; when the proposal was submitted in 1990, these were the only sites under serious consideration. At NSF's urging, site consideration expanded to include Mauna Kea and the Atacama Desert in Chile. In May 1998, an NRAO study strongly recommended that the MMA be built in the Atacama Desert [6].

Large Southern Array[3]

In the late 1980s, discussions took place in Europe regarding a possible millimeter array to be built in the Southern Hemisphere. A European study group was formed, and the Large Southern Array (LSA) project began at a meeting in 1991 as a proposal for an array with a total collection area of 10,000 m². In 1994, a recommendation made by the ESO millimeter working group to establish a permanent millimeter advisory committee was endorsed. Later that year, the group proposed that a design study be initiated. In April 1995, a memorandum of understanding concerning a study for a large millimeter array in the Southern Hemisphere was signed by the ESO, the Institut de Radio Astronomie Millimétrique, the Onsala Space Observatory, and the Netherlands Foundation for Research in Astronomy. The group began to develop antenna concepts and performed detailed testing at several sites in Chile [9].

ALMA

Because similar sites were being examined for the LSA and the MMA in northern Chile, the possibility of a partnership became obvious. In June 1997, an agreement was signed by the ESO and the NRAO to explore such a partnership. The agreement established three joint working groups: a Science Working Group to consider the scientific objectives, a Technical Working Group, and a Management Working Group [9]. The LSA and the MMA had different concepts and requirements, which were reconciled after detailed study of four antennae [9]. The two projects officially merged in June 1999 to become the Atacama Large Millimeter Array [1]. The current ALMA design has 64 12-m antennae with a total collecting area of some 7,000 m².

ALMA first received MREFC funding for design and development work in FY 1998. An antenna prototype began testing in New Mexico in the fall of 2002 [11]. In February 2003, Rita R. Colwell (director of NSF) and Catherine Cesarsky (director general of the ESO) signed an agreement to jointly construct and operate ALMA [10]. The first ALMA production antenna will be delivered to Chile in FY 2006 [11], early science observing will begin in 2007, and full-scale operations in FY 2012.

---

[3]Japanese radioastronomers have also been developing a Large Millimeter-Submillimeter Array (LMSA). The possibility of merging the LMSA and MMA has been discussed since 1995, but no decisions have been made [5].

References

[1] Background Information: Atacama Large Millimeter Array. National Radio Astronomy Observatory.

[2] Robert P. Chase. 1990. Millimeter Astronomers Push for New Telescope. *Science* 249:1504.

[3] Alan H. Barrett. Report of Subcommittee on Millimeter- and Submillimeter-Wavelength Astronomy. April 1983 Astronomy Advisory Committee, National Science Foundation; MMA Memo No. 9.

[4] M. Mitchell Waldrop. 1983. Astronomers Ponder a Catch-22. *Science* 220:698.

[5] Al Wootten. Historical Information About the MMA. Jan. 25, 1999. Available at <http://www.cv.nrao.edu/~awootten/mmaimcal/mmahistory.html>.

[6] Recommended Site for the Millimeter Array. 1998. National Radio Astronomy Observatory, May.

[7] Frazer Owen. Interoffice Memo. 1982.The Concept of a Millimeter Array. Very Large Array, National Radio Astronomy Observatory, September 10. MMA Memo No. 1.

[8] Paul A. Vanden Bout, director, NRAO. 1997. FY98 Budget – National Science Foundation, Subcommittee on Basic Research, House Committee on Science, April 9.

[9] LSA/MMA Feasibility Study, April 1998. Available at <http://www.eso.org/projects/alma/doclib/reports/lsa_report98/report_june99.html>.

[10] Charles E. Blue and Richard West. 2003. U.S. and European ALMA Partners Sign Agreement. National Radio Astronomy Observatory, Press Release, Feb. 25. Available at <http://www.nrao.edu/pr/2003/almasigning/index-p.shtml>.

[11] Major Research Equipment and Facilities Construction. National Science Foundation Fiscal Year 2004 Budget Request.

[12] S. Radford and L. Nyman. 2001. ALMA Project Book, Version 5.5, Chapter 14; July 25. Chajnantor Site Studies: Overview available at <http://www.tuc.nrao.edu/mma/sites>.

[13] National Research Council. 1991.The Decade of Discovery in Astronomy and Astrophysics. Washington, D.C.: National Academy Press.

[14] Correspondence from NSF, October 2003.

[15] Approved Minutes of 367th NSB Meeting (NSB 02-53), March 14, 2002.

EARTHSCOPE

Description

EarthScope, a geographically distributed geophysical and geodetic instrument array, will seek to deploy a large and diverse array of instrumentation over North America to learn "how the continent was put together, how it is moving now, and what is beneath it" [1]. EarthScope will comprise the US Seismic Array (USArray), the Plate Boundary Observatory (PBO), the San Andreas Fault Observatory at Depth (SAFOD), and the satellite-based Interferometric Synthetic Aperture Radar (InSAR). The first three will be funded through the NSF MREFC account, and the latter is planned to be jointly developed with the National Aeronautics and Space Administration (NASA). US Array and SAFOD are referred to as phase I, and PBO as phase II. See Table C-2 for a timeline of the major developments.

TABLE C-2 Timeline of Major Developments

Before 1998	Discussions among members of earth-sciences community to identify facilities needs for future research [2].
1998-2002	Concept development for EarthScope continues with NSF R&RA funding [3].
July 1999	EarthScope presented to NSB Committee on Programs and Plans for FY 2001 budget planning.
October 3-5 1999	Workshop on PBO in Snowbird, Utah [18].
November 30 - December 1, 1999	EarthScope identified as long-term GEO funding need during fall NSF Advisory Committee for Geosciences meeting [4].
May 1-2, 2000	GEO/AC announces $17.44 million request for NSF MREFC account in FY 2001 to fund EarthScope [5].
October 2000	EarthScope listed as part of tools development in NSF GPRA strategic plan for FY 2001-FY 2006 [6].
October 30 - November 1, 2000	Second PBO workshop in Palm Springs, California [18].
May 3-4, 2001	USArray Design Workshop in San Diego, California [18].
May 22-25, 2001	PBO workshop in Pasadena, California [18].
September 6, 2001	NSF director Rita R. Colwell identifies EarthScope as among top funding priorities [8].
September 7, 2001	NSF director receives letter from president of Geological Society of America encouraging placement of EarthScope high on MREFC priority list [9].
October 2001	NSB identifies EarthScope as among highest priorities.
October 10-12, 2001	EarthScope workshop in Snowbird, Utah [18].
October 29, 2001	National Research Council review of EarthScope integrated science [10].
December 11, 2001	USArray Steering Committee meeting in San Francisco, California [18].
January 30 - February 1, 2002	EarthScope education and outreach workshop in Boulder, Colorado [18].
February 4, 2002	President Bush signs budget proposal for FY 2003, including $35 million for EarthScope [11].
February 2002	Earth-sciences community launches letter-writing campaign to ensure approval of EarthScope funding [11].
February 11-13, 2002	USArray Steering Committee meeting in Washington, D.C. [18].

continued

TABLE C-2 Continued

March 2002	US-Canada PBO workshop in Seattle, Washington [18].
March 25-27, 2002	EarthScope information-technology workshop in Snowbird, Utah, results in formation of EarthScope Information Technology Forum [14].
June 2002	US-Mexico PBO workshop in San Diego, California [18].
June 12, 2002	Drilling for pilot hole into San Andreas fault for SAFOD begins with funding from International Continental Drilling Program [19].
August 4-7, 2002	Creating the EarthScope Legacy workshop in Snowbird, Utah [18].
October 31 - November 3, 2002	EarthScope workshop on active magmatic systems in Vancouver, Washington [18].
November 26, 2002	EarthScope Science and Education Committee (ESEC) formed [15]
December 6-10, 2002	American Geophysical Union Special Session on EarthScope in San Francisco, California [18].
January 10-11, 2003	ESEC meeting in Washington, D.C. [18].
February 3, 2003	President Bush's budget proposal for FY 2004 includes $45 million for EarthScope [16].
February 20, 2003	President Bush signs budget for FY 2003, allocating $30 million for EarthScope [12,13].
March 2-4, 2003	EarthScope Complementary Geophysics workshop in Denver, Colorado [18].
April 17, 2003	NSF releases solicitation for science and education proposals for EarthScope [20].
April 23-25, 2003	USArray and the Great Plains meeting in Manhattan, Kansas [18].
June 15, 2003	House budget proposal includes $43.5 million for EarthScope in FY 2004 [17].

Approval and Funding History

Funding for construction of USArray and PBO was requested in FY 2001, but Congress did not provide it. EarthScope was included in the draft FY 2002 request but was not included in the request to Congress. The project (all three elements) was included in the FY 2003 request to Congress and was funded.

Managing Institutions

Incorporated Research Institutions for Seismology will manage USArray, UNAVCO, Inc., PBO, and Stanford University SAFOD.

Development Summary

Initial discussions concerning what would become EarthScope began over a decade ago when earth scientists began identifying observations and measurements needed to address natural hazards and to answer outstanding problems in the earth sciences. Through a series of NSF-funded workshops and conferences, a cohesive set of planning documents outlining specific needs for large observational facilities emerged. That also helped to establish a precedent for cooperation between the scientific community and government agencies including NSF, the US Geological Survey, and NASA. By the late 1990s, the various concepts were consolidated into the single EarthScope initiative [2]. Initial funding for concept development was provided by the NSF Research and Related Activities (R&RA) account from FY 1998 to FY 2002 [3].

During the fall 1999 meeting of the NSF Advisory Committee for Geosciences (GEO/AC), EarthScope was identified as one of the long-term funding needs for the NSF Division of Earth Sciences (EAR) [4]. During the spring 2000 meeting, Margaret Leinen, assistant director of the NSF Directorate for the Geosciences (GEO), announced that $17.44 million for FY 2001 was requested from Congress for the NSF MREFC account to initiate construction of USArray and SAFOD [5]. The request was denied, and funding for EarthScope development continued through the NSF R&RA account [3].

In the 2001-2006 NSF Government Performance and Results Act (GPRA) strategic plan submitted in October 2000, NSF identified the development of "Tools—Broadly accessible, state-of-the-art information bases and shared research and education tools" [6] as one of its three overarching goals. EarthScope was listed as part of the tools-development plan, and it was highlighted as one of two new programs for investment in tools by NSF Director Dr. Rita R. Colwell at the February 7, 2000, FY 2001 NSF budget briefing [7].

On September 6, 2001, Colwell identified EarthScope as one of three top NSF MREFC account priorities [8]. On the following day, she received a letter from the president of the Geological Society of America encouraging NSF to make EarthScope a top-priority MREFC request [9]. In October 2001, at its 365th meeting, the National Science Board (NSB) approved Resolution NSB 01-180 indicating that EarthScope was among the board's highest priorities. Also in October 2001, the National Research Council published a review of the EarthScope facility development and science program. Conducted at the request of NSF, the review bolstered support for the program. The Research Council found "that EarthScope is an extremely well articulated project that has resulted from consideration by many scientists over several years, in some cases up to decades . . . The committee conclude[d] that EarthScope will have a substantial impact on earth science in America and worldwide" [10].

On February 4, 2002 the president signed the FY 2003 budget request, including $35 million for the NSF MREFC account for EarthScope. The budget did not include funding for some MREFC projects for which funding had previously been requested. Anticipating a difficult battle with the Senate Appropriations Committee for approval of the EarthScope portion of the budget, the earth-sciences community launched a congressional letter-writing campaign [11]. On February 20, 2003, President Bush signed into law the Omnibus Spending Bill for FY 2003. The bill had been passed by both houses of Congress a week before that and included $30 million for EarthScope [12, 13].

After the March 2002 EarthScope Information Technology Workshop, the EarthScope Information Technology Forum (ESIT) was formed. The ESIT aims include coordinating current EarthScope information technology (IT) developing future EarthScope IT, and standardizing data structures and interfaces [14].

On November 26, 2002, EAR and the earth-sciences community formed the EarthScope Science and Education Committee (ESEC). Pursuant to a recommendation made by the National Research Council (2001), the ESEC will "provide leadership and a central point-of-contact for the major elements of the EarthScope project . . . [and] serve as a conduit for information between the funding agencies and the scientific communities" [15].

On February 3, 2003, President Bush submitted his FY 2004 budget request, including $45 million for EarthScope [16]. On July 15, 2003, the House Subcommittee on Veterans Affairs [VA], Housing and Urban Development [HUD], and Independent Agencies of the Committee on Appropriations' version of the budget indicated funding for EarthScope below the requested level, at $43.5 million [17].

References

[1] National Research Council. 2001. Review of EarthScope Integrated Science. Washington, D.C: National Academy Press.
[2] EarthScope history. Available at <dax.geo.arizona.edu/earthscope/about/history.html>.
[3] EarthScope funding profile.
[4] NSF GEO Advisory Committee November 30–December 1, 1999 meeting minutes.
[5] NSF GEO Advisory Committee May 1-2, 2000 meeting minutes.
[6] NSF GPRA Strategic Plan FY 2001–FY 2006, October 3, 2000.

[7] Remarks by NSF Director Rita Colwell at NSF FY 2001 budget briefing.
[8] Testimony of NSF director before House Committee on Science Subcommittee on Research, September 6, 2001.
[9] Letter from GSA President to NSF Director, September 7, 2001.
[10] National Research Council. Review of EarthScope Integrated Science. Washington, D.C.: National Academy Press, 2001.
[11] EarthScope News Release, February 4, 2002.
[12] California Department of Education Federal Update, February 14, 2003.
[13] EarthScope News Release, February 20, 2003.
[14] EarthScope News Release, April 12, 2003.
[15] EarthScope News Release, November 26, 2003.
[16] EarthScope News Release, February 3, 2003.
[17] NSF OLPA Congressional Update, July 15, 2003.
[18] EarthScope past meetings available at <www.earthscope.org/news/past_mtgs.html>.
[19] EarthScope News Release, June 12, 2002.
[20] NSF Progam Solicitation for EarthScope: Science, Education and Related Activities for the USArray, San Andreas Fault Observatory at Depth (SAFOD) and Plate Boundary Observatory (PBO), April 17, 2003.

GEMINI OBSERVATORIES

Description

The Gemini Observatory is a new generation of twin optical infrared telescopes that use innovative instruments and new observational and operational approaches. It consists of two 8.1-m telescopes that are sensitive to optical and infrared light. Gemini North sits atop Mauna Kea in Hawaii on a 2-acre site subleased from the University of Hawaii (UH) [1]. Gemini South is at Cerro Pachon in the Chilean Alps on land held by the Association of Universities for Research in Astronomy (AURA) [2]. Together, the telescopes provide an unprecedented opportunity for studying the entire northern and southern sky. The project is an international collaboration of seven nations: the United States, the UK, Canada, Australia, Chile, Brazil, and Argentina. NSF, which contributes 50 percent of the funding for Gemini, serves as the executive agency for the project. AURA serves as the project's managing body. The National Optical Astronomy Observatory (NOAO) acts as the gateway for US involvement in Gemini. See Table C-3 for a timeline of the major developments. Dedicated in 1999, Gemini North made news with its first data release, providing dramatic images of the galactic center. The project's construction phase ended in January 2002 with the dedication of Gemini South.

Approval and Funding History

MREFC funding for construction began in FY 1995. Construction was initiated in FY 1991.

TABLE C-3 Timeline of Major Developments

Before 1992	Gemini operated under directorship of NOAO [3].
Spring 1992	US Gemini involvement under AURA management independent of NOAO [3].
April 16, 1993	AURA board recommends creation of US Gemini Office in NOAO [4].
1993	UK and Canada join United States in providing funding for Gemini [6].
October 6, 1994	Groundbreaking ceremony for Gemini North [15].
October 22, 1994	Groundbreaking ceremony for Gemini South [15].
1995	Creation of MREFC account; Gemini receives $41 million [7].
October 11, 1995	Corning Inc., announces completion of Gemini North primary mirror blank [10].
February 8, 1996	Fred Gillett named International Gemini Project scientist [16].
May 1, 1997	Corning Inc., completes fabrication of Gemini South mirror blank [11].
June 24, 1997	Groundbreaking ceremonies at Hilo Base Facility for Gemini North [12].
July 29, 1997	Chile rejoins Gemini partnership [17].
February 18, 1998	Australia joins Gemini partnership [18].
June 28, 1998	Primary mirror delivered to Mauna Kea site of Gemini North [19].
November 18, 1998	Dedication ceremony for Hilo Base Facility [20].
April 15, 1999	Gemini receives NSF grant for improved Internet access to Gemini North [12].
June 25, 1999	Dedication ceremony for Gemini North [13].
March 17, 2000	Primary mirror delivered to Cerro Pachon site of Gemini South [21].
October 16, 2000	Gemini North penetrates into galactic core with first data release [13].
January 7, 2002	Gemini North data reveal Brown Dwarf orbiting a Sun-like star [22].
January 18, 2002	Dedication ceremony for Gemini South [14].
November 13, 2002	Gemini North named in honor of Fred Gillett [23].

Managing Institutions

The project is governed by the Gemini board, which was established by the Gemini agreement signed by the participating agencies. NSF is the executive agency for the seven-nation partnership and acts on its behalf.

Development Summary

The US Gemini project began under the directorship of the NOAO to "expedite the start-up phase" [3]. In the spring of 1992, because of its international nature, it became an independent project subject to rules different from those set forth for the NOAO by an NSF-AURA collaborative agreement [3]. In April 1993, AURA established a US Gemini Office in the NOAO to act as "the focus for US involvement in the international Gemini project" [4]. In contrast with other NOAO undertakings, however, the US Gemini Office does not operate Gemini; rather, it serves as the avenue for facilitating US research and development related to Gemini [5].

During calendar 1991 and 1992, the United States provided the sole financial contributions to Gemini in an amount totaling $12.1 million [6]. Those funds were allocated from the NSF R&RA account [7]; the MREFC account had not yet been established. UK and Canadian involvement began in 1993, but the United States remained the primary contributor, providing about 60 percent of the funding [6]. In FY 1995, with the establishment of the NSF MREFC account, the United States contributed $41 million to Gemini [6], about one-third of the total MREFC budget [8]. Except for $4 million in FY 1998, all later NSF funding for Gemini operations has been allocated through the R&RA account. US financial involvement exceeded its 50 percent partnership in the early years of the project, but it will be reimbursed by other member countries. All member countries' contributions will match their proportion of the partnership by 2005 [9].

In October 1994, groundbreaking ceremonies marked the beginning of construction on both the Gemini North and Gemini South telescopes. Poor weather during that Hawaiian winter, however, delayed work on Gemini North, pushing back "first light" for the northern telescope by 5 months [6]. Parallel to the onsite construction, Corning Inc. began manufacture of the 8.1-m primary mirrors; it completed the first blank in October 1995 [10]. The mirror blank for Gemini South was completed in May 1997 [11].

In June 1997, groundbreaking ceremonies took place at UH at Hilo University Park, the site of the sea-level operations for Gemini North. The facility was completed in November 1998. In spring of 1999, Gemini received a $600,000 grant from NSF to increase Internet access to and between Gemini North facilities. Coupled with a $350,000 grant given to

UH's Information Technology Services, the grant allowed high-speed access to Gemini's large, high-resolution data files and better community outreach via the Internet [12].

The June 1999 dedication of Gemini North ushered in a new era of optical astronomy for a new millennium. The first data release, in October 2000, provided a spectacular glimpse into the core of the Milky Way [13]. The dedication ceremony for Gemini South occurred in January 2002, for the first time allowing complete coverage of the entire sky from an 8-m-class observatory [14]. Since the completion of both telescopes, Gemini data have yielded a steady stream of scientific papers.

References

[1] Gemini public webpage at <http://www.gemini.edu/public/mauna.html>.
[2] Gemini public webpage at <http://www.gemini.edu/public/pachon.html>.
[3] Gemini Project Newsletter, Number 1, March 1992.
[4] NOAO Newsletter No. 34, June 1, 1993.
[5] Gemini Project Newsletter No. 5, June 1993.
[6] The International Gemini Telescopes Annual Report, 1995.
[7] Gemini Funding Profile.
[8] Frontiers, September 1998.
[9] The International Gemini Telescopes Annual Report, 1998, p. 35.
[10] Corning News Release, October 11, 1995.
[11] Corning News Release, May 9, 1997.
[12] Gemini Observatory Press Release, April 15, 1999.
[13] Gemini Observatory Press Release, October 16, 2000.
[14] Gemini Observatory Press Release, December 3, 2001.
[15] Gemini Observatory Press Release, September 30, 1994.
[16] Gemini Observatory Press Release, February 8, 1996.
[17] Letter from NSF staff associate for Gemini, August 29, 1997.
[18] Gemini Observatory Press Release, February 18, 1998.
[19] Gemini Observatory Press Release, June 29, 1998.
[20] Gemini Observatory Press Release, November 18, 1998.
[21] Gemini Observatory Press Release, March 20, 2000.
[22] Gemini Observatory Press Release, January 7, 2002.
[23] Gemini Observatory Press Release, November 13, 2002.

HIAPER (HIGH-PERFORMANCE INSTRUMENTED AIRBORNE PLATFORM FOR ENVIRONMENTAL RESEARCH)

Description

HIAPER is a jet aircraft with unique high-altitude and advanced payload research capabilities. It is used for research on Earth systems, including atmospheric and weather research, on regional and planetary scales. HIAPER is a middle-size jet research aircraft capable of carrying up to

6,000 lb of payload to an altitude of 51,000 ft with a range of 6,000 nautical miles. Those features will enable atmospheric studies in and near the tropopause and allow for long-range studies of coastlines and borders. The MREFC portion of the project involves acquisition and modification of a Gulfstream GV airframe and the development of advanced instrumentation that will match platform capabilities and scientific needs. The changes made to the airframe include installing hard points under the wings for carrying pods or sensors, optical ports for remote-sensing equipment, and additional hard points for attaching various other apparatus to the aircraft. In addition, this GV will differ from those commercially available for private use in having a much sparser interior to allow for more instrumentation and advanced cyberinfrastructure to support the needs of researchers. In providing a state-of-the-art facility for airborne atmospheric studies, HIAPER heralds unprecedented advances in the field.

Approval and Funding History

Approved in August 1998, MREFC funding for HIAPER began in FY 2000. It included support for planning, design, and development and for acquisition and modification of the airframe. NSF's FY 2002 budget request did not include MREFC funding for HIAPER. The House Subcommittee on VA, HUD, and Independent Agencies of the Committee on Appropriations added $35 million for HIAPER; this funding was retained in the final spending bill for FY 2002. The final $25.536 million for HIAPER construction was appropriated in FY 2003.

Managing Institutions

The National Center for Atmospheric Research (NCAR), operated by the 69-member university consortium called the University Corporation for Atmospheric Research (UCAR), and NSF acquired the HIAPER airframe from Gulfstream Corporation. Lockheed Martin is contracted to modify the aircraft structurally to meet scientific requirements. Once modifications are completed, the NSF-owned aircraft will be operated and maintained by the NCAR Atmospheric Technology Division (ATD).

Development History

The need for a middle-size jet research aircraft was identified and repeatedly reiterated at a series of workshops beginning in 1982. See Table C-4 for a timeline of the major developments. The outcome of meetings in 1982, 1987, and 1992 was nearly unanimous support from NSF-funded scientists for listing the acquisition of such a vehicle for the NSF research-

TABLE C-4 Timeline of Major Developments

1980s	Workshops identify the need for new HIAPER aircraft [1].
1989	Mid-Sized Jet Review Committee strongly endorses need for new middle-size jet [1].
1997	NSF GEO proposes HIAPER [1].
August 1997	NSB approves HIAPER project development plan [2].
1998	HIAPER receives first R&RA money for concept development [3].
Summer 1998	Survey conducted to obtain input into HIAPER from university community [6].
August 1998	NSB authorizes NSF to seek MREFC funding for HIAPER in FY 2000 [5].
May 24-26, 1999	HIAPER community workshop in Boulder, Colorado [14].
2000	HIAPER receives first MREFC funding.
June 19, 2000	UCAR issues request for proposal to purchase HIAPER aircraft [7].
Fall 2001	Completion of negotiations with Gulfstream for purchase of aircraft; NSB approves resulting deal [8].
December 2001	Contract awarded to Gulfstream for production of HIAPER "green" airframe [8].
June 2002	Completion of airframe and transfer to Lockheed Martin [9].
November 4-6, 2002	HIAPER instrumentation workshop [10].
Early 2003	Preliminary design review for HIAPER modifications [11].
May 14-15, 2003	Technical interchange meeting between Gulfstream and Lockheed Martin HIAPER teams and NCAR and UCAR staff [11].
June 24-26, 2003	Critical design review [12].

aircraft fleet as the highest priority. A 1989 ad hoc committee convened to examine NCAR-prepared reports on the scientific justification of such a project, the Mid-Sized Jet Review Committee, strongly endorsed the need [1]. The specifications identified for such an aircraft at those meetings included payload, altitude, and range, all of which represented vast technologic improvements over the existing research fleet [1].

In light of the continued support for acquiring a new aircraft, the NSF Directorate for the Geosciences (GEO) proposed HIAPER in FY 1997 [1]. The NSB approved the project development plan in August 1997 [2].

HIAPER received NSF R&RA money during FY 1998 and FY 1999 to fund concept development [3].

In August 1998, the NSB authorized NSF to seek MREFC funding for HIAPER in FY 2000. After that confidence-boosting development, NCAR's ATD successfully submitted a proposal to carry out HIAPER on NSF's behalf [4, 5]. During the summer of 1998, a survey of potential users was conducted to elicit input from the university community [6]. In May 1999, a workshop held in Boulder, Colorado, developed more specific requirements for the aircraft.

In June 2000, UCAR issued a request for proposals for the purchase of a HIAPER aircraft [7]. During that year, HIAPER received its first MREFC funding, and additional funding was appropriated in FY 2001. During the ensuing 2 years, Gulfstream was selected as a potential contractor, and terms were negotiated. The negotiations concluded in fall 2001 and received NSB approval. A contract with Gulfstream for the purchase of a GV airframe was signed in December 2001. The FY 2002 MREFC budget included $35 million for the purchase of the aircraft [8].

The airframe was completed and delivered to UCAR in June 2002 [9]. It was then transferred to Lockheed Martin in Greenville, South Carolina, for design and implementation of the modifications necessary to support scientific missions. In November 2002, a workshop was held in Boulder, Colorado, to discuss instrumentation needs for HIAPER [10].

In early 2003, NCAR, UCAR, NSF, and the HIAPER Advisory Committee participated in a preliminary design review in preparation for the June critical design review (CDR) [11]. The CDR revealed "no major surprises," leaving Lockheed Martin ready to begin making modifications in August 2003 [12]. Modification work on HIAPER is expected to reach completion in Autumn 2004, when the aircraft will be transferred to Boulder, Colorado, for complete testing, interior modification, and deployment of cyberinfrastructure necessary to support future research missions. The first science mission is expected to take place in 2005 [10].

References

[1] Background of the HIAPER Initiative, available at <www.hiaper.ucar.edu>.

[2] Minutes from UCAR Board of Trustees meeting, October 6-7, 1997.

[3] HIAPER funding profile.

[4] UCAR Staff Notes Monthly. HIAPER: A research plane for the 2000s takes shape, November 1998.

[5] UCAR Quarterly. New aircraft is cleared for takeoff, Winter 1998.

[6] UCAR Quarterly. Under construction: the new NSF/NCAR high-altitude jet, Spring 2002.

[7] UCAR RFP No. 18883.

[8] UCAR Staff Notes Monthly. Full speed ahead for new NSF/NCAR high-altitude jet, April 2002.

[9] What is HIAPER? Available at <www.hiaper.ucar.edu>.
[10] UCAR Quarterly. HIAPER instrument workshop set for November, Fall 2002.
[11] HIAPER Project Office Monthly Update, May 2003.
[12] HIAPER Project Office Monthly Update, July 2003.
[13] Senate Report 108-143, September 2003.
[14] Agenda for HIAPER Community Workshop in Boulder, Colo., May 1999.

ICECUBE

Description

IceCube will be a kilometer-size neutrino detector built in the Antarctic ice. The Antarctic Muon and Neutrino Detector Array (AMANDA) and AMANDA-II demonstrated the feasibility of using the clarity and depth of the ice sheet at the South Pole to detect high-energy neutrinos. See Table C-5 for a timeline of the major developments. With a 1-km^3 volume sitting almost 2.5 km below the surface, the world's largest telescope to date will detect neutrinos traveling through the earth's core from the northern sky with a pointing accuracy of about 1 degree. Building on the state-of-the-art drilling and detector techniques developed for the earlier projects, IceCube will revolutionize the field of particle astrophysics by allowing the detection of PeV-energy neutrinos [1]. Those neutrinos are believed to result from some of the highest-energy processes observed in the universe, such as gamma-ray bursts and active galactic nuclei. The completion of IceCube will enable us to look deeper than ever before into formation processes and other previously opaque cosmic events.

Approval and Funding History

IceCube was approved by the NSB in October 2000 for submission in a future budget request. MREFC funding was first included in the President's budget request to Congress in FY 2004. However, Congress provided funding for Ice Cube startup activities in FY 2002 and FY 2003 in the amounts of $15 million and $24.7 million, respectively. The FY 2004 NSF budget request to Congress included $60 million for the start of full construction. Both House and Senate markups have included funding for start of construction, albeit at lower levels than the FY 2004 request. (As of this writing, the conference committee has not met, so the FY 2004 appropriation has not been made.) The total cost of the construction, including the startup funding mentioned above, is projected to be $251.6 million. US funding for the project is provided to the University of Wisconsin-Madison and then to subawardee institutions and to the support contractors of NSF's US Antarctic Program (USAP)-Raytheon Polar

TABLE C-5 Timeline of Major Developments

1987	Francis Halzen proposes Antarctic ice as a site for neutrino detector [10].
1992	Construction begins on AMANDA [10].
1996	AMANDA observes atmospheric neutrino candidates [1].
November 1999	IceCube proposal received from University of Wisconsin for full construction funding; mail review takes place in March 2000; no funding action taken.
January 2000	Completion of AMANDA-II [1].
March 2000	Proposal reviewed by NSF-DOE Scientific Assessment Group for Experimental Non-Accelerator Physics.
June 2000	Halzen gives talk at International Symposium on High Energy Gamma-Ray Astronomy proposing IceCube [1]; NSF conducts full IceCube project baseline review.
October 2000	NSB approves IceCube for inclusion in the FY 2002 or later budget [NSB-00-165].
October 2001	NSF conducts second full IceCube baseline review.
November 6, 2001	House and Senate approve FY 2002 funding of $15 million for IceCube startup activities [2].
December 2001	University of Wisconsin submits proposal for IceCube startup phase in amount of $15 million.
January 2002	HEPAP's Panel on Long Range Planning endorses IceCube [4].
March 2002	NSB approves award of $15 million for IceCube startup activities.
March 29, 2002	OSTP requests National Research Council study on neutrino projects [6].
December 2002	National Research Council report finds no redundancy between IceCube and underground science laboratory [7].
February 3, 2003	President includes funding for IceCube in FY 2004 budget [8].
February 20, 2003	Omnibus Spending Bill passed by Congress (PL 108-7) provides $24.7 million for continued for IceCube startup work. Because law also provides for recision of 0.65 percent, amount available for IceCube became $24.5 million.
May 28, 2003	NSB approves up to $24.5 million for continuation of IceCube startup activities [9].

Services Company and the US Air National Guard (for flight services to the South Pole Station).[4]

Managing Institutions

The work will be undertaken by the IceCube Collaboration, led by the University of Wisconsin (UW). Of the 11 US institutions in the collaboration, 10 are universities and the other is the Lawrence Berkeley National Laboratory. The collaboration also includes foreign participants representing three countries—Belgium, Germany, and Sweden—contributing an estimated $40 million toward the cost of construction.

Development History

The 1996 success of AMANDA and completion of AMANDA-II in 2000 demonstrated the feasibility of using Antarctic ice as a neutrino detector. The expandable photomultiplier-tube technology developed for AMANDA helped to pave the way for larger-scale Antarctic detectors. A full proposal for IceCube funding was submitted in November 1999 and was later reviewed for scientific merit and technical readiness. In June 2000, Francis Halzen, the AMANDA principal investigator, gave a talk at the International Symposium on High Energy Gamma-Ray Astronomy that indicated readiness to begin work on a 1-km^3 neutrino detector called IceCube. On the basis of the project's intellectual merit and state of planning, the NSB in October 2000 approved the inclusion of funding for IceCube construction in the NSF FY 2002 budget request. In 2001, UW continued developmental activities for the project, and NSF conducted a further baseline review.

In November of 2001, the $15 million was included in the joint House-Senate appropriations bill FY 2002 MREFC funding for startup activities associated with the successor IceCube project [2].

In its January 2002 report, the Department of Energy–NSF High-Energy Physics Advisory Panel (HEPAP) Subpanel on Long Range Planning for High-Energy Physics endorsed IceCube as part of the US high-energy physics roadmap [4]. After the release of the funding in March 2002, the UW-based project began hiring engineers and administrators [5]. Although the president approved the FY 2002 funding for IceCube, concerns regarding the redundancy of various neutrino-detector projects [6] prompted the White House Office of Science and Technology Policy (OSTP) to request a study by the National Research Council to review

---

[4]The USAP facilities include Amundsen-Scott South Pole Station, McMurdo Station, Palmer Station, and two research vessels [3].

neutrino projects that were under way. The White House was concerned about possible overlap of science goals between IceCube and an underground science laboratory [6]. The 2003 National Research Council report found no substantial overlap between the two projects, noting that they enabled essentially different types of neutrino detection. IceCube is optimized to detect high-energy neutrinos, and the underground science laboratory offers a low-background environment for studying lower-energy neutrinos [7].

The key IceCube startup activities include development and production of in-ice devices (the photodetectors at the heart of IceCube), development of an enhanced hot-water drill to drill the 2500-m-deep holes in the ice cap into which the photodetectors will be deployed, and data systems development for acquisition, transmission, archiving, and analysis of the data from the roughly 5,000 distinct photosensors in the IceCube array. Drilling and deployment of the IceCube sensors is expected to take six austral summer seasons; completion is estimated in FY 2010.

The president's FY 2004 budget included funding for IceCube at the level of $295.2 million through FY 2013 [8]. In May 2003, the NSB approved up to $25 million for UW and the USAP to complete the phase 1 effort to develop the hot-water drill and its associated support equipment and to commence developing a design for the downhole electronics modules that operate the photodetectors.

References

[1] Francis Halzen. High Energy Neutrino Astronomy: Towards Kilometer-Scale Detectors, astro-ph/0103195v1, March 13, 2001.
[2] John Fauber. Milwaukee Journal Sentinel, November 7, 2001.
[3] Funding Profile for IceCube available at <www.nsf.gov>.
[4] DOE/NSF HEPAP Subpanel on LRP for US HEP report. The Science Ahead: The Way to Discovery, Particle Physics in the 21st Century, January 28, 2002.
[5] Ernie Mastroianni. Milwaukee Journal Sentinel, April 1, 2001.
[6] *Nature* 417:5, May 2, 2002.
[7] National Research Council. Neutrinos and Beyond: New Windows on Nature. Washington, D.C.: The National Academies Press, 2003.
[8] IceCube Press Update dated August 7, 2003, available at <icecube.wisc.edu>.
[9] NSF Media Advisory (NSB 03-77), May 28, 2003.
[10] *The Baltimore Sun*, July 21, 2003.

IODP (INTEGRATED OCEAN DRILLING PROGRAM)

Description

The Integrated Ocean Drilling Program (IODP) will be a multivessel, multinational project to drill and take cores in oceanic settings to "investigate a wide range of earth system processes" [1]. See Table C-6 for a timeline of the major developments. A successor to the single-vessel, 18-year, international Ocean Drilling Program (ODP), the IODP will take advantage of new technologies to enable a variety of ocean-drilling studies. The current project plan entails two ships funded by the United States and Japan. The Japanese contribution will be a riser ship, named *Chikyu*, capable of drilling to 7 km below the seafloor in 4-km-deep waters, far enough to reach the earth's mantle. The US ship will be similar to the existing ODP vessel but with "significantly enhanced coring and drilling capabilities" [2]. European countries have expressed interest in providing smaller, mission-specific platforms (MSPs) for the IODP. Beyond the construction phase, the IODP will enable further cooperation in international ocean-drilling research. The annual operating costs will be shared by member countries that pay for "IODP participation units." The United States and Japan will each pay for one-third of the units, and the remaining one-third will be divided among member countries. Each unit of participation provides the member country with representation on drilling cruises and the science advisory panel [2]. MSP and nonriser operations are scheduled to begin in 2004, and riser and nonriser operations are scheduled to start in late 2006.

Approval and Funding History

The IODP has not yet received MREFC funding. It was included in the NSF FY 2004 budget proposal as an out-year request for funding in FY 2005.

Managing Institutions

The project is managed through a memorandum of cooperation between NSF and the Japanese Ministry of Education, Culture, Sports, Science and Technology (MEXT). NSF has entered into a contract with Joint Oceanographic Institutions, Inc. (JOI), a 10-institution nonprofit consortium for management and operations of the nonriser vessel.

Development Summary

The IODP builds on 35 years of successful scientific ocean drilling that began with the 15-year Deep Sea Drilling Project (DSDP) initiated in 1968 [1, 3]. After the retirement of the DSDP vessel *Glomar Challenger*, the US drill ship *JOIDES Resolution* set sail to inaugurate the ODP in January 1985 [4]. The ODP, an NSF project funded through JOI, was an international endeavor ultimately involving scientists from over 20 nations [5].

TABLE C-6 Timeline of Major Developments

1968	*Glomar Challenger* sets sail for the Deep Sea Drilling Project [3].
January 22, 1985	*JOIDES Resolution* sets sail for the Ocean Drilling Project [4].
1993	US Committee to Consider Post-1998 Scientific Ocean Drilling (COMPOST-I) report [8].
1996	Ocean Drilling Program Long Range Planning Committee report [8].
Early 1997	NSF requests USSAC study to assess degree of US commitment to ocean drilling beyond 2003 [8].
February 16-17, 1997	US Committee to Consider Post-2003 Scientific Ocean Drilling (COMPOST-II) meets at University of Miami [8].
March 6-7, 1997	Review of COMPOST-II draft report [8].
Mid 1997	Conference on Cooperative Ocean Riser Drilling in Tokyo, Japan [9].
1999	Formation of IODP Planning Subcommittee [10].
May 26-29, 1999	Conference on Multiple Platform Exploration in Vancouver, Canada [10, 11].
1999	Formation of USSAC Conceptual Design Committee [10].
March 2000	Delivery of Conceptual Design Committee report to NSF [13].
July 12, 2001	NSF director testifies at hearing on ocean exploration and ocean observatories before House Committees on Resources and Science [5].
September 2001	JOI/USSAC report on US participation in IODP [14].
January 18, 2002	*Chikyu* launching ceremony and announcement of UK participation in IODP [15].
June 12-14, 2002	Conference on US participation in IODP held in Washington, D.C. [1].
2003	IODP included in out-year MREFC funding request for FY 2005 in NSF budget proposal.
March 19, 2003	NSF solicits US contractor to manage scientific and drilling operations of nonriser vessel [17].
April 22, 2003	United States and Japan sign memorandum of cooperation for IODP [17].
August 4, 2003	NSF issues solicitation for US science-support program associated with IODP [18].

The ODP is overseen by JOI for Deep Earth Sampling (JOIDES). Originally planned as a 10-year project, the *JOIDES Resolution* set sail on its final leg in July 2003 after 18 years of successful operations. The IODP is scheduled to begin the next phase of scientific ocean drilling on October 1, 2003 [6].

In 1997, pending the scheduled completion of the ODP, NSF requested a study from the US Science Advisory Committee (USSAC) to "assess the degree of US commitment to scientific ocean drilling beyond 2003" [7]. The USSAC then formed the US Committee to Consider Post-2003 Scientific Ocean Drilling (COMPOST-II), the successor to COMPOST-I, formed in 1993 to consider post-1998 scientific drilling [7, 8]. Drawing on the 1993 National Research Council report *Solid-Earth Sciences and Society* and the 1996 ODP Long Range Planning Committee report, COMPOST-II made six recommendations affirming "its commitment to a new international scientific ocean drilling program post-2003" [8]. The report endorsed many of the elements incorporated in the IODP, including the multi-platform approach. *IODP* appeared as a working title for the project in 1997.

In July 1997, during the Conference on Cooperative Ocean Riser Drilling held in Tokyo, Japan, 150 scientists and engineers met to discuss scientific study enabled by a riser-equipped drilling vessel. Their recommendations were sent to the International Working Group for the IODP (IWG/IODP), cochaired by officials of NSF and the Japanese Science and Technology Agency [9].

Planning on the IODP continued to move quickly through the late 1990s. An IODP Planning Subcommittee (IPSC) of JOIDES was formed in 1999 [10]. In May of that year, the Conference on Multiple Platform Exploration (COMPLEX), held in Vancouver, Canada, drew over 350 participants [10, 11]. COMPLEX set out to "define the 'intellectual challenges' of the post-2003 scientific ocean drilling program" [12]. The results of the conference laid out an ambitious scientific agenda for nonriser drilling research and helped to form the basis of recommendations on the nonriser IODP vessel. NSF's charging the USSAC with the conceptual-design task for the nonriser drill ship led to the formation of the USSAC Conceptual Design Committee [10]. By spring 2000, the conceptual-design report was delivered to NSF. By then, NSF had made clear its intention to make an award to acquire and modify or convert a nonriser drill ship in October 2003 if funds were available [13]. Funding for the IODP at sub-million-dollar levels from the NSF R&RA account began in FY 2000.

In July 2001, at a hearing before the House Committees on Resources and Science, NSF Director Rita R. Colwell discussed the IODP as the "future phase of scientific drilling" [5]. In September of that year, the JOI/USSAC published *Understanding Our Planet Through Ocean Drilling:*

A Report from the United States Science Advisory Committee, which made the case for US participation in the IODP [1]. A shorter version of the report intended for a wider audience, *United States Participation in the Integrated Ocean Drilling Program 2003-2013*, endorsed US participation in the IODP to "ensure that the best science is pursued, innovation is incorporated, technical operations run smoothly, and scientific exploration culminates in synthesis and integration" [14, 1].

In January 2002, a launching ceremony was held for *Chikyu*, the Japanese riser drill ship. At the ceremony, NSF Deputy Director Joseph Bordogna announced UK support for the IODP and anticipated the coming formal agreements between the United States and Japan on its management [15].

To provide background for a written recommendation to NSF regarding the IODP, a conference on US participation was held in June 2002. The conference examined management and structural strategies for the US component of the IODP [1]. That fall, the NSF Division of Ocean Sciences reported "the expectation that the long-term responsibility for ODP scientific and physical assets will be transferred to appropriate contractors and organizations in the planned follow-on program, the Integrated Ocean Drilling Program (IODP) as it is developed and implemented" [16].

On April 22, 2003, NSF and MEXT signed a formal memorandum of cooperation for the IODP. A month before the signing, NSF issued an NSB-approved solicitation for a US contractor to manage the scientific and drilling operations of the US nonriser vessel. Contract negotiations were expected to be completed in August 2003 [17]. Also in August, NSF issued a solicitation for "a qualified provider to facilitate and enhance the participation of the US scientific community in all aspects of IODP" [18]. The IODP was included in the NSF FY 2004 budget as an out-year request for MREFC funding in FY 2005.

References

[1] John Farrell. IODP Planning Status. JOI/USSAC Newsletter, Summer 2002.
[2] JOI/USSAC report. United States Participation in the Integrated Ocean Drilling Program, September 2001.
[3] Kasey White. ODP: International Earth Science. Geotimes, August 2001.
[4] Ocean Drilling Program News Release, January 22, 1985.
[5] Testimony of Rita R. Colwell, NSF director, before the House Committees on Resources and Science at Hearing on Ocean Exploration and Ocean Observatories, July 12, 2001.
[6] Ocean Drilling Program News Release, Leg 210, Summer 2003.
[7] NSF OCE Newsletter, Spring 1998.
[8] COMPOST-II Report. A New Vision for Scientific Ocean Drilling, 1997.
[9] NSF OCE Newsletter, Fall 1997.
[10] NSF OCE Newsletter, Fall 1999.
[11] COMPLEX Report, May 1999.

[12] <http://www.oceandrilling.org/COMPLEX/Default.html>.
[13] NSF OCE Newsletter, Spring 2000.
[14] JOI/USSAC Report. United States Participation in the Integrated Ocean Drilling Program, September 2001.
[15] NSF OLPA release of Dr. Bordogna's remarks at *Chikyu* launching ceremony in Kobe, Japan, January 18, 2002.
[16] NSF OCE Newsletter, Fall 2002.
[17] NSF Press Release (NSF PR 03-41), April 22, 2003.
[18] Program Solicitation (NSF 03-586) US Science Support Program Associated with the Integrated Ocean Drilling Program (USSSP-IODP), August 4, 2003.

LHC (LARGE HADRON COLLIDER)

Description

The Large Hadron Collider (LHC) will be the world's highest-energy accelerator facility. The United States is involved in construction of the LHC accelerator and two particle detectors, ATLAS and CMS. The LHC is a high-energy particle-physics facility designed to collide protons at teravolt (TeV) energies. The LHC, in Geneva at the European Laboratory for Particle Physics (CERN), is one of the largest international scientific enterprises yet undertaken. LHC participants include the 20 member states of CERN,[5] the United States, Canada, India, Russia, Japan, and physicists of many other countries. Designed to fit inside the tunnel constructed for CERN's Large Electron Positron Collider (LEP), the LHC heralds a new age in high-energy physics. By providing a 10-fold increase in energy and a 1,000-fold increase in intensity over current colliders, the LHC will enable scientists "to study the collisions of the tiny quarks locked deep inside protons [1]," an order of magnitude smaller than has been studied until now. Over 1,000 superconducting magnets, cooled to temperatures below that of outer space and sustaining a magnetic field more than 16,000 times that of Earth, will accelerate the protons to the necessary energies [1]. In addition to the magnets, precision detectors able to withstand high levels of radiation must be developed and built to "observe" the collision products. Two large detectors—a toroidal LHC apparatus (ATLAS) and the compact muon solenoid (CMS)—are key elements of the LHC project and involve the collaborative efforts of more than 4,000 people in 45 countries [1]. The funding for US participation in the LHC comes from two sources: the Department of Energy (DOE) and NSF. DOE's Brookhaven National Laboratory (BNL), Lawrence Berkeley National Labora-

---

[5]Member states of CERN: Austria, Belgium, Bulgaria, Czech Republic, Denmark, Finland, France, Germany, Greece, Hungary, Italy, Netherlands, Norway, Poland, Portugal, Slovak Republic, Spain, Sweden, Switzerland, and UK.

tory, and Fermi National Accelerator Laboratory will carry out the US role in magnet design and manufacture. DOE and NSF funds will contribute to ATLAS and CMS, efforts that involve more than 550 scientists at nearly 60 universities and six DOE national laboratories [1].

Approval and Funding History

Construction began with MREFC support in FY 1999.

Managing Institutions

The LHC is an international project under construction at the CERN Laboratory in Geneva, Switzerland. NSF has awarded grants to Northeastern University and Columbia University under cooperative agreements, with subcontracts to over 50 US universities. A total of 34 international funding agencies participate in the ATLAS detector project and 31 in the CMS detector project.

Development Summary

Discussions about the LEP began in the late 1970s. See Table C-7 for a timeline of the major developments. Because of the high cost of excavating the LEP tunnel, scientists at CERN decided to begin looking into possible next-generation accelerators to replace LEP at the same site. A 1984 joint European Committee for Future Accelerators (ECFA) and CERN workshop recommended exploring the TeV range for future colliders. The following year saw the formation of the CERN Long Range Planning Committee, which recommended installing a multi-TeV facility in the tunnel after the completion of the LEP program. Development of a proposal for such a facility continued throughout the late 1980s and resulted in a plan approved by the CERN Scientific Policy Committee in 1990 [2].

In late 1991, the CERN Council agreed in a unanimous decision that the LHC was "the right machine for the further significant advance in the field of high energy physics research and for the future of CERN" [3]. The Council then asked for a full technical, scientific, and financial proposal by 1993. The LHC External Review Committee endorsed the resulting proposal in December 1993 [2]. The Council, however, determined that the cost associated with meeting the target completion date of 2002 exceeded the CERN budget. Discussion ensued regarding the possibility of seeking contributions from nonmember states [2].

In the United States, 1993 saw another important event in the history of particle physics: the cancellation of the Superconducting Super Collider (SSC) project by Congress. The SSC had represented the future of US

TABLE C-7 Timeline of Major Developments

1977	Preparatory discussions for LEP raise possibility of next-generation collider [2].
March 1984	ECFA-CERN workshop recommends exploring TeV-range facility [2].
1985	CERN Long Range Planning Committee proposes installing 1-TeV facility in LEP tunnel after project's completion [2].
June 1990	CERN Scientific Policy Committee endorses proposal for 1-TeV facility [2].
Late 1990	ECFA backs 1-TeV facility proposal [2].
December 1991	CERN Council approves pursuing LHC and requests full proposal by 1993 [3].
October 1993	Cancellation of US SSC project; subpanel of HEPAP formed to examine future options for US high-energy physics [4].
December 1993	LHC External Review Committee endorses LHC proposal [2].
December 17, 1993	CERN Council hears proposal and costs for LHC and encourages contributions from nonmember states [2].
May 23, 1994	US subpanel recommends participation in LHC [4].
July 1994	Secretary of Energy recommends US participation in LHC to Congress [4].
1994	Canada and Japan consider entry into LHC [4].
December 16, 1994	CERN Council approves two-stage plan for LHC with possibility of expediting project with outside contributions [5].
May 10, 1995	Japan announces 5-billion-yen contribution to LHC accelerator [6].
December 15, 1995	CERN director general announces that ATLAS and CMS have passed peer review and are pending approval [7].
1996	Canada contributes Can $30 million to LHC accelerator [8].
1996	NSF funds development work for LHC through R&RA account [10].
January 8-9, 1996	CERN director general leads delegation to Washington, D.C., to begin negotiations for US involvement in LHC [7, 8].
March 1996	India contributes US $12.5 million to LHC [8].
June 1996	Russia contributes 67 million Swiss francs to LHC [8].
July 1996	United States announces tentative plans to contribute $531 million to LHC accelerator and detectors [9].

TABLE C-7 Continued

August 1996	CERN member states propose reduced contributions to annual budget [8].
December 20, 1996	CERN Council decides to pursue LHC as single-stage project to be completed in 2005; Council decides to reduce contributions from member states without altering LHC budget; Japan announces second contribution of 3.85 billion yen to LHC accelerator [8].
May 15, 1997	Chairman of House Committee on Science notes progress toward addressing his concerns with DOE's proposal for LHC appropriations [11].
December 8, 1997	United States signs agreement to contribute $531 million to LHC accelerator and detectors [12].
May 7, 1998	NSB approves NSF participation in LHC [17].
May 18, 1998	Japan contributes additional 5 billion yen to LHC accelerator.
August 8, 1998	French government approves commencement of LHC civil engineering [13].
1999	First year of NSF MREFC funding for LHC [10].
December 13, 1999	NSF and DOE sign memorandum of understanding outlining US management role in LHC [18].
November 8, 2000	LEP shuts down, making way for LHC [19].
January 21, 2003	US delivers first components for LHC [15].
June 20, 2003	CERN Council confirms schedule for LHC start in 2007 [16].

particle physics, and its cancellation led to the formation of a subpanel of the US High Energy Physics Advisory Panel (HEPAP) charged with examining the possibility of international particle-physics collaborations. The subpanel's report, released in May 1994, recommended that the United States join the LHC project with a potential contribution of about $400 million for the LHC accelerator and detectors. Those recommendations were taken to Congress in July 1994. During this time, Canada and Japan also began considering entry into the LHC project [4].

After deliberating costs and schedules, the CERN Council approved construction of the LHC in December 1994. Outside contributions had not yet been secured, but the Council decided on a two-stage construction process to be completed in 2008 using only funds from member states. The resolution also noted that CERN welcomed contributions from nonmember states toward the LHC and would re-examine the possibility of a single-stage project if sufficient funds materialized [5].

Once CERN had announced its definite intention to proceed with the LHC, Japan promised a contribution of 5 billion yen to the LHC accelerator in May 1995 [6]. During January 1996, CERN's director general led a delegation to Washington, D.C., to begin negotiations concerning a US role in the LHC project [7, 8]. In July 1996, NSF announced tentative plans, subject to congressional approval, for US involvement in the LHC project totaling $531 million, with $81 million to come from NSF [14]. During that year, CERN reached agreements for contributions to the LHC from India, Russia, and Canada [8], and NSF began funding LHC-related development through its R&RA account [10]. With the influx of outside funding, the CERN Council decided to proceed with the single-stage plan for the LHC [8].

During the annual CERN Council meeting in December 1996, member states decided to reduce their annual contributions to CERN, although the director general said that this would not alter the amount of resources devoted to the LHC [8]. The reduction in funds, along with concerns about the US management role in the project, raised objections in Congress. DOE began working with the House Committee on Science to rectify those issues [11]. US officials signed an agreement with CERN in December 1997, promising to contribute $531 million to the LHC project, of which $81 million would come from the NSF MREFC account for ATLAS and CMS [12]. Civil-engineering work for the LHC began in the following summer [13].

NSF MREFC funds were assigned to the LHC from FY 1999 to FY 2003 [10]. The LEP project reached completion in 2000 [14]. In 2003, the United States delivered its first piece of hardware to Geneva, one of 20 25-ton superconducting magnets to be built at BNL [15]. The LHC is on schedule to begin full operations in April 2007 [16].

References

[1] CERN Press Release (PR07.97), August 12, 1997.
[2] CERN Press Release (PR08.94), The Large Hadron Collider, June 17, 1994.
[3] CERN Press Release (PR12.93), December 17, 1993.
[4] CERN Press Release (PR06.94), LHC, A World Project, June 17, 1994.
[5] CERN Press Release (PR16.94), December 16, 1994.
[6] CERN Press Release (PR05.95), May 12, 1995.
[7] CERN Press Release (PR11.95), December 15, 1995.
[8] CERN Press Release (PR09.96), December 20, 1996.
[9] Frontiers: NSF Electronic Newsletter, July/August 1996.
[10] Large Hadron Collider Funding Profile.
[11] NSF Office of Legislative and Public Affairs Congressional Update, May 15, 1997.
[12] CERN Press Release (PR07.97), December 8, 1997.

[13] CERN Press Release (PR04.98), August 10, 1998.
[14] CERN Press Release (PR17.00), December 15, 2000.
[15] CERN Press Release (PR01.03), January 21, 2003.
[16] CERN Press Release (PR08.03), June 20, 2003.
[17] Approved Minutes of 342nd NSB Meeting (NSB 97-96), May 7, 1998.
[18] NSF OLPA News Tip, December 20, 1999.
[19] CERN Press Release (PR14.00), November 8, 2000.
[20] CERN Press Release (PR02.98), May 18, 1998.

LIGO (LASER INTERFEROMETER GRAVITATIONAL-WAVE OBSERVATORY)

Description

The Laser Interferometer Gravitational-Wave Observatory (LIGO) comprises two sites: one in Livingston, Louisiana, and one in Hanford, Washington. Both sites have L-shaped Michelson interferometers, with 4-km arms, that are designed to detect the extremely tiny (10^{-19} m) differential stretching of space caused by the passage of gravitational waves (GWs) (the Hanford site also has a second interferometer with 2-km arms housed in the same structure). Because the effect is expected to be so small, two geographically separated sites are necessary to eliminate local sources of noise that can mimic a GW signal. Bona fide signals must have common characteristics in all three interferometers and be observed nearly simultaneously at both sites. If successful, LIGO could open a new avenue of astronomy: GW astronomy. Several foreign groups are working on similar, but smaller, observatories. Collaborative data sharing between LIGO and the other groups will allow refinements in the identification of GW signals and enhance the precision with which astrophysical sources can be identified.

Approval and Funding History

Construction began in FY 1992 and was supported by R&RA funds from the NSF Directorate for Mathematical and Physical Sciences (MPS). MREFC funding of construction began in FY 1994. Civil construction was completed in FY 1998. LIGO was commissioned in FY 2001 and began scientific operations in FY 2002.

Managing Institutions

LIGO is managed by the California Institute of Technology (Caltech) under a cooperative agreement with NSF. A memorandum of understanding with Caltech makes the Massachusetts Institute of Technology

(MIT) a full partner in the design, construction, and operation of the LIGO facilities.

Development Summary

The concept of detecting GWs with laser interferometers was first developed in the 1960s. Table C-8 presents a timeline of the major developments. The work of Rainer Weiss (MIT), Ronald Drever (University of Glasgow, UK), and Kip Thorne (Caltech) in the 1960s and 1970s was of particular importance for the development of LIGO. Weiss demonstrated a laser interferometer in 1967 and conceived of using it for GW detection. Using military funding, he began construction of a 1.5-m prototype at MIT. Drever, with James Hough, began constructing a 10-m prototype interferometer in 1975 at Glasgow. In 1968, Thorne began a theoretical effort to study GWs and their sources. After being convinced that GWs could be detected, he prompted Caltech to initiate an experimental effort in 1978. Drever took the lead of Caltech's experimental effort in 1979, splitting his time evenly between Caltech and Glasgow.

In 1975, NSF began providing funds to MIT to continue work on the 1.5-m prototype interferometer. In 1979, after a review by an NSF subcommittee under R. Deslattes of the work of MIT and Caltech, the NSF Physics Division Advisory Committee endorsed the concept of a GW interferometer. In 1980, NSF provided funds to MIT to complete the 1.5-m prototype interferometer and a technical site and cost study of a large-baseline interferometer. It also funded a 40-m prototype interferometer, which Drever and Stan Whitcomb began constructing at Caltech in 1981. The Caltech interferometer began running in July 1982 and became a testbed for the future LIGO design. In 1983, the MIT study concluded that it would be technologically feasible for a 1-km-scale interferometer to detect GWs. Under pressure from NSF, which did not want to fund two separate GW projects, Caltech and MIT joined forces in 1983 to create plans for the LIGO project. Under their agreement, Caltech had the lead administrative role, but the LIGO Steering Committee consisted of three partners: Weiss, Thorne, and Drever (at Caltech).

Late in 1983, Caltech and MIT scientists made a joint presentation to the NSF Physics Advisory Committee. The committee gave LIGO second priority after improvements at the Cornell Synchrotron. In 1984, a memorandum of understanding was signed between Caltech and MIT for joint design and construction of LIGO. Frank Schutz (Jet Propulsion Laboratory, JPL) was appointed project manager and initiated studies of possible sites for LIGO's two interferometer facilities. In March 1984, the LIGO Steering Committee made a presentation to NSB.

In 1986, the International Society of General Relativity and Gravita-

TABLE C-8 Timeline of Major Developments

1960	Joseph Weber initiates work on GW detectors by using narrow-band acoustic bars at University of Maryland [16].
1962	Gertsenshtein and Pustovoit (G&P) in USSR conceive basic idea of laser interferometer GW detector (*Soviet Physics – JETP*, 14, 84) [16].
Late 1960s-1972	Weber, independently of G&P, suggests interferometer detector to Robert Forward at Hughes Aircraft; Forward, with Moss and Miller at Hughes, builds prototype interferometer and runs it as readout for GW detector [16].
1967-mid 1970s	In 1967, Rainer Weiss (MIT) demonstrates photon-shot noise-limited laser interferometer; independently of other groups, Weiss conceives idea of interferometer's use for GW detection and initiates detailed analysis of it and developmental research; in 1972, he publishes his analysis, identifying all fundamental noise sources that such an interferometer must face, and conceives ways to deal with them[8]; he initiates construction of 1.5-m prototype; all this is done with military funding but is terminated by Mansfield Amendment before prototype is operational [16].
1973	Hans Billing, having worked on Weber-type bar detectors, initiates interferometer development at Max Planck Institute for Quantum Optics in Garching, Germany; this research program ultimately leads in 1990s to German component of GEO600 project and in 2000s to German contribution to Advanced LIGO [16].
1975	NSF begins funding work of Rainer Weiss at MIT; Ronald Drever and James Hough, having worked on Weber-type bar detectors, initiate interferometer development at University of Glasgow, UK; they begin construction of 10-m prototype interferometer [16].
1968-79	Kip Thorne in 1968 creates theoretical effort on GWs and their sources at Caltech and, convinced that GW detection will succeed, triggers Caltech in 1978 to initiate experimental GW research; in 1979, Drever accepts offer to lead Caltech's experimental effort and splits his time between Caltech and Glasgow (until 1984, when he moves full-time to Caltech) [16].
April 1979 [1, 4]	NSF Division of Physics Advisory Committee (triggered by Weiss's MIT work and Drever's new program at Caltech) endorses concept of GW interferometer [7, 16].

continued

---

[8]Quarterly Progress Report of the Research Laboratory of Electronics, MIT, **105**, 54.

TABLE C-8 Continued

1980	NSF funds completion of MIT's 1.5-m prototype interferometer [16]; NSF funds MIT for technical, site, and cost study of large-baseline interferometer system (Paul Linsay, Peter Saulson, Rainer Weiss), which is planned to be basis of multiuniversity consortium proposal to NSF [16]; NSF funds construction of 40-m prototype interferometer at Caltech [16].
1981	Ron Drever and Stan Whitcomb initiate construction of 40-m prototype interferometer [16].
June 1982 [16]	40-m interferometer achieves first lock, becoming (like Glasgow, Garching, and MIT prototype interferometers) a testbed for future large-baseline LIGO and GEO interferometers.
1983	MIT study is completed (with input from Arthur D. Little and Stone & Webster Engineering Corp) and concludes that 1-km-scale interferometers with adequate sensitivity to detect cosmic GWs are technologically feasible [4, 16]; under pressure from NSF, which was concerned about available manpower, Caltech (Drever, Thorne) agrees to join forces with MIT (Weiss) to create LIGO project for constructing two 1-km interferometer facilities; plans for LIGO are based on results of MIT study and on experimental work at Caltech, MIT, Glasgow, and Garching; Caltech takes lead administrative role in LIGO; LIGO Steering Committee (Drever, Thorne, and Weiss) is formed [4, 7, 16].
Late 1983	Caltech and MIT make joint presentation to NSF Division of Physics Advisory committee, which ranks LIGO second to improvements at Cornell Synchrotron and above University of Illinois Microtron [16].
1984	Memorandum of understanding between Caltech and MIT for joint design and construction of LIGO; project manager (Frank Schutz, JPL) is appointed and initiates studies of possible sites for LIGO's two interferometer facilities [16].
March 1984 [1, 16]	Steering Committee presentation to NSB enables NSF (Isaacson, Bardon) to encourage Caltech and MIT to submit proposal for final design study for LIGO.
November 1984 [1]	NSB approves development plan for LIGO.
December 1984 [16]	Caltech and MIT jointly submit proposal to NSF for final design study for LIGO; proposal is turned down largely for financial reasons.
1985	Caltech and MIT submit revised proposal for final design study; proposal is turned down, this time both for financial reasons and because some referees do not deem project ready for final design study [16].
1986	National Research Council Physics Survey endorses LIGO [8].

TABLE C-8 Continued

November 1986 [1,16]	NSF (Marcel Bardon) appoints blue-ribbon panel to examine LIGO's scientific case, technical feasibility, management plan, and costing; panel, cochaired by Boyce McDaniel (Cornell) and Andrew Sessler (Lawrence Berkeley National Laboratory), includes university and industrial scientists and engineers, among them Garwin; after week-long review, panel enthusiastically endorses LIGO's scientific case and technical feasibility but not its management; panel insists that LIGO Steering Committee (which was unable to make technical decisions on rapid timescales required for large project) be replaced by single director. [4, 7].
1987	LIGO Steering Committee gives strong support to UK and Germany to build 3-km equilateral-triangle interferometer detector in Bavaria; this project becomes mired in costs of unification of Germany but is predecessor of current GEO600 in Hanover.
June 1987 [1]	Rochus Vogt (Caltech) is appointed LIGO project director and principal investigator [3, 7, 13], and LIGO Steering Committee is disbanded [16].
December 1987 [16]	Caltech and MIT submit joint proposal under Vogt's leadership for 3 years of R&D, which will lead to submission of LIGO construction proposal.
February 1988 [1]	NSF review by panel of experts and site visit.
1988	LIGO R&D proposal is funded.
October 1988 [1]	Presentations on LIGO to NSF's Division of Physics and Division of Astronomy Advisory Committees.
December 1989 [1, 7, 16]	Caltech and MIT, under Vogt's leadership, submit LIGO construction proposal to NSF [15].
February 1990 [1]	NSF review of LIGO construction proposal and site visit.
April 1990 [1]	NSB approves LIGO construction proposal.
Fall 1990	NSF requests and Congress rejects LIGO construction funding for FY 1991 [3, 7].
November 1990	NSB approves site-selection process.
April 1991 [6]	Hearing on LIGO before House Committee on Science, Space, and Technology—witnesses Vogt (LIGO), Clifford M. Will (McDonnell Center for the Space Sciences, Washington University), and Tyson (AT&T Bell Labs); Will is strongly supportive; Tyson is not [16].
April 1991 [5]	Caltech announces list of 18 proposed sites for LIGO.

continued

TABLE C-8 Continued

May 1991 [9]	Rep. Rick Boucher (D-VA) notes that LIGO is absent from National Academy of Sciences Astronomy and Astrophysics Survey report; Bahcall explains that it is in NAS Physics Survey.
Fall 1991	Congress approves first year of funding for LIGO ($23 million) [2].
Winter 1991	Congress funds LIGO for first year [2].
November 1991[1]	NSF review by panel of experts validates site-evaluation process; comprehensive evaluation of all sites sent to NSF.
February 1992 [2, 10]	NSF announces selection of two sites: Hanford and Livingston.
May 1992 [1]	Cooperative agreement is signed by Caltech and NSF.
July 6 1992 [10]	LIGO is restructured with Drever no longer a direct participant.
November 1992 [1]	NSF review by panel of experts recommends "dedicated" NSF program manager.
February 1993 [1]	David Berley is appointed NSF program manager for LIGO.
April 1993 [1]	Berley forms LIGO Coordinating Group.
June 1993 [1]	NSF review by panel of experts supports NSF concerns about LIGO project management.
December 1993 [11]	NSF freezes spending on construction-related contract until Vogt comes up with acceptable management plan, including how to accommodate outside scientists.
January 1994 [11]	Congress, citing NSF's management concerns with LIGO, tells NSF to cut $8 million from LIGO budget.
January 1994 [1, 11, 16]	After consultation with relevant NSF personnel, with LIGO's scientific leaders, and with MIT's president, president of Caltech reaches decision to replace LIGO Director Vogt.
February 1994 [1, 11]	Barry Barish (Caltech, formerly with SSC) is appointed laboratory director by president of Caltech in consultation with NSF and MIT [16]; he hires Gary Sanders (formerly with SSC) as project manager [13].
June 1994 [1]	NSF LIGO cost review by expert panel, which recommends that LIGO increase contingency.
July 1994 [1, 2]	Groundbreaking at Hanford.
November 1994 [1]	Project-management plan approved by NSF.
July 1995 [2]	Groundbreaking at Livingston.

TABLE C-8 Continued

October 1996 [1]	NSF review panel recommends formation of LIGO Scientific Collaboration; LIGO laboratory director selects R. Weiss as first spokesperson of LIGO Scientific Collaboration [14].
August 1997 [14]	First meeting of LIGO Scientific Collaboration.
November 1999 [2]	LIGO inauguration ceremony.
August 2002 [2]	First scientific operation of all three interferometers.

tion endorsed LIGO. Shortly afterward, Marcel Bardon (NSF) appointed a blue-ribbon panel to examine LIGO's scientific justification, technical feasibility, management plan, and budget. The panel, chaired by Boyce McDaniel (Cornell) and Andrew Sessler (Lawrence Berkeley National Laboratory), included university and industrial scientists and engineers, among them Richard Garwin. After a week-long review, the panel enthusiastically endorsed LIGO's scientific justification and technical feasibility but not its management plan. The panel insisted that a single director replace the LIGO Steering Committee. Also in 1986, the National Research Council Physics Survey endorsed LIGO. In July 1987, Rochus Vogt (Caltech) was appointed project director and principal investigator, and the LIGO Steering Committee was disbanded.

Under Vogt's leadership, Caltech and MIT received 3 years of R&D funding beginning in 1988; that led to the submission of a LIGO construction proposal to NSF in December 1989. The proposal received NSB approval. LIGO was included in the FY 1991 budget request, but the funding was denied in Fall 1990. In May 1991, during discussions of amendments to the NSF Authorization Act, Rep. Rick Boucher (D-Va.) noted the absence of LIGO in the NRC Astronomy and Astrophysics Survey Report. Ultimately, Congress approved first-year funding for LIGO in the fall of 1991. In February 1992, NSF announced the selection of Hanford and Livingston[6] as the two LIGO sites.

Over the next 2 years, LIGO and NSF faced several management and organizational issues. In May 1992, Caltech and NSF signed a cooperative agreement. In February 1993, after a recommendation by a panel of experts to form a "dedicated" LIGO management position in NSF, David

---

[6]Hanford was in the district of the speaker of the House, Tom Foley (D-WA), and Livingston was in the state of Sen. J. Bennett Johnston (D-LA), who sat on the Senate committee that appropriates money for NSF.

Berley was appointed program manager for LIGO, and the LIGO Coordinating Group (LCG) was created.[7] NSF, in carrying out its oversight responsibilities, expressed serious concern that the LIGO project needed to be restructured before it could effectively move forward. That concern was shared by Congress. In December 1993, NSF froze spending on a construction-related contract pending submission of an acceptable management plan that included how to accommodate outside scientists. Congress, after briefing by NSF staff, requested that NSF delay $8 million from the LIGO budget in January 1994. As a consequence of NSF's concern and congressional intervention, the management structure of LIGO was reconstituted in February 1994 with Barry Barish (Caltech) as new principal investigator and Gary Sanders (Caltech) as new project manager. By the end of the year, with new management in place and a satisfactory project-management plan approved by NSF, construction began. In November 1994, LIGO's new management presented revised costs for strengthening project management to the NSB, which approved the increase.

In 1997, as LIGO neared the end of its construction phase, two organizational institutions were formed [14]: LIGO Laboratory and LIGO Scientific Collaboration. LIGO Laboratory consists of the facilities supported by NSF under LIGO Operations and Advanced R&D; this includes administration of the LIGO detector facilities and the support and test facilities at Caltech, MIT, Hanford, and Livingston.

LIGO Scientific Collaboration [14] is a forum for organizing technical and scientific research in LIGO. Its mission is to ensure equal scientific opportunity for individual participants and institutions by organizing research, publications, and all other scientific activities. It includes scientists from LIGO Laboratory and collaborating institutions. It is a separate organization from LIGO Laboratory with its own leadership and governance, but it reports to the LIGO Laboratory Directorate for final approval of its research program, technical projects, observational physics publications, and talks announcing new observations and physics results.

LIGO was inaugurated in November 1999, when construction activities were substantially complete. Since then, the project has carried out commissioning activities interleaved with progressively more sensitive data gathering. Those undertakings involve the LIGO Scientific Collaboration (a group of more than 45 institutions and 450 scientists) in the scientific activities of LIGO.

---

[7]LCG consists of members in the NSF Office of Budget and Finance Award Management, Division of Grants and Agreements, Office of the General Council, Office of Legislative and Public Affairs, and Mathematical and Physical Sciences Directorate [1].

References

[1] Victor Cook (NSF). NSF Management and Oversight of LIGO. Large Facility Projects Best Practices Workshop (NSF), Sept. 21, 2001.

[2] LIGO chronology. LIGO Press & Media Kit. Available at <http://www.ligo.caltech.edu/ LIGO_web/PR/scripts/chrono.html>.

[3] Jeffrey Mervis. Funding of two science labs revives pork barrel vs. peer review debate. *The Scientist* 5[23]:0, Nov. 25, 1991.

[4] Robert Buderi. Going after gravity: How a high-risk project got funded. *The Scientist* 2[17]:1.

[5] Malcolm W. Brown. Experts clash over project to detect gravity wave. *New York Times*, April 30, 1991, p. C1.

[6] FY 1992-FY 1993 National Science Foundation authorization. Hearing before the House Committee on Science, Space, and Technology, March 13, 1991. CIS-NO: 91-H701-51.

[7] M. Mitchell Waldrop. Of politics, pulsars, death spirals—and LIGO. *Science* 249:1106-1108.

[8] NRC Physics Survey Committee, Physics Through the 1990s: A Summary. Washington, D.C.: National Academy Press, 1986.

[9] Rick Boucher. Introduction of National Science Foundation authorization act amendments of 1991. Congressional Record, Vol. 137, No. 70, May 9, 1991.

[10] John Travis. LIGO: A $250 million gamble. *Science* 260:612-614.

[11] Christopher Anderson. LIGO director out in shakeup. *Science* 263:1366.

[12] William T. Broad. Big science squeezes small-scale researchers. *New York Times*, December 29, 1992, p. C1.

[13] Robert Irion. LIGO's mission of gravity. *Science* 288:420-423.

[14] LIGO Scientific Collaboration (LSC) available at <www.ligo.org>. (Official LIGO website: <www.ligo.caltech.edu>.)

[15] W. Wayt Gibbs. Ripples in spacetime. *Scientific American* 28:62.

[16] Correspondence with Barry Barish, who consulted with Rainer Weiss and Kip Thorne.

NEES (GEORGE E. BROWN, JR. NETWORK FOR EARTHQUAKE ENGINEERING SIMULATION)

Description

The George E. Brown, Jr. Network for Earthquake Engineering Simulation (NEES) will be a geographically distributed national network of shared experimental earthquake engineering research equipment sites linked by a high-performance Internet system; it will consist of three major components:

• Next-generation earthquake engineering research equipment (such as shake tables, a tsunami wave basin, geotechnical centrifuges, large-scale laboratory facilities, and mobile and permanently installed field equipment) distributed around the country at 15 universities.

• NEESgrid, a high-performance network that will connect the remote sites and enable remote equipment operation and experimental viewing, distributed experimentation, collaboration, data sharing, and simulation via the Internet.

- A nonprofit university consortium, NEES Consortium, Inc., which will be responsible for NEES operation and management during FY 2005– FY 2014.

Approval and Funding History

NEES was approved in November 1998 by the NSB for possible inclusion in the FY 2000 budget. MREFC funding began in FY 2000 for design, development, and construction.

Managing Institutions

To construct NEES, NSF has made 18 awards: 16 awards for equipment to 15 institutions, one award to the University of Illinois at Urbana Champaign for network system integration through NEESgrid, and one award to the nonprofit Consortium of Universities for Research in Earthquake Engineering (CUREE) for development of the nonprofit university consortium.

Development Summary

In October 1994, 9 months after the Northridge, California, earthquake, the National Earthquake Hazards Reduction Program (NEHRP) Reauthorization Act (PL 103-374) was signed into law. See Table C-9 for a timeline of the major developments.

As part of the act, the President was required to "conduct an assessment of earthquake engineering research and testing capabilities." To comply, NSF and the National Institute of Standards and Technology (NIST) cosponsored an assessment study by the Earthquake Engineering Research Institute (EERI). A workshop, attended by 65 invited participants, was the primary element of the study and resulted in a frequently cited report released in September 1995 [1]. The report recommended an increase in funding and support for earthquake research, as many reports had in the past. In October, Daniel P. Abrams, chair of the EERI study, testified before the House Committee on Science.

In December 1995, NSF held a small workshop on the future of earthquake engineering experimental research to develop an action plan [8]. The National Network for High Performance Seismic Simulation (NHPS,

TABLE C-9 Timeline of Major Developments

January 17, 1994	Northridge, California, earthquake.
October 19, 1994	NEHRP Reauthorization Act is signed into law (PL 103-374); it requires earthquake engineering research assessment.
January 17, 1995	Kobe, Japan, earthquake.
May 1995	EERI begins assessment study sponsored by NSF and NIST [1].
July 31-August 1, 1995	EERI holds workshop in San Francisco as primary element of assessment; 65 invited participants attend [1].
September 1995	EERI releases report [1]: *Assessment of Earthquake Engineering Research and Testing Capabilities in the United States.*
October 23, 1995	Abrams testifies on EERI report before House Committee on Science [15, 4].
December 1995	NSF workshop on future directions in earthquake engineering experimental research [8].
June 1997	NSF internal project-development plan for NHPS.
Late summer 1997	NSB gives go-ahead to ENG to develop detailed project-management proposal [6].
October 1, 1997	1997 NEHRP Reauthorization Act is signed into law (PL 105-47).
October 7, 1997	NSF announces (NSF PR 97-59) new funding for three centers (University of California, Berkeley; University of Illinois, Urbana-Champaign; and State University of New York, Buffalo) to form consortia for earthquake engineering [12].
October 15, 1997	EERI meeting in San Francisco, California, directed by Jim Jirsa (University of Texas), to assist NSF in planning for NHPS [6].
February 1998	NHPS session is held at EERI meeting in San Francisco [7].
February 2-April 1998	At EERI experimental-research forum, it is unanimously decided that NHRP should be managed by consortium of universities [7].
May 8-9, 1998	NSF-sponsored NHPS workshop: tsunami and coastal engineering and research community, in Baltimore, Maryland (18 participants) [9].
May 28-29, 1998	NSF-sponsored NHPS workshop: geotechnical earthquake engineering research community at University of California, Davis (40 participants) [2a, 2b].
June 1998	Eugene Wong, information-technology specialist formerly with OSTP, takes over as assistant director of ENG [16, 12].
June 4, 1998	NHPS meeting in Seattle, Washington, arranged by Jirsa and Abrams after sixth national conference on earthquake engineering, for potential consortium members [7].

continued

TABLE C-9 Continued

November 1998	NSB approves inclusion of NEES in NSF FY 2000 budget [10].
February 1999	NEES included in NSF FY 2000 budget [12].
December 1999	Solicitation is issued (NSF 00-6) for NEES earthquake engineering research equipment, phase 1 [12, 13]; solicitation is issued (NSF 00-7) for NEES system integration [12, 14].
February 10, 2000	NSF holds roundtable discussion with presenters from geosciences, earthquake engineering, and computer science; purpose is to address what is expected and required of NEES [4].
September 2000	Award (NSF 00-7) to University of Illinois, Urbana-Champaign, for earthquake engineering research community workshop held November 16-17, 2000 [12].
October 2000-January 2001	In award (NSF 00-6), NSF makes 11 awards to 10 universities for new equipment and upgrades totaling $45 million [12, 13].
January 2001	Solicitation is issued (NSF 01-56) for NEES consortium development [12].
August 2001	Award (NSF 00-7) to University of Illinois, Urbana-Champaign, for NEESgrid ($10 million) [12, 14].
September 2001	Solicitation is issued (NSF 01-164) for NEES earthquake engineering research equipment, phase 2 [12].
October 2001	Award (NSF 01-56) to CUREE for NEES consortium development project ($2 million) [12].
February-September 2002	21 regional workshops are held by NEES consortium development project [11, 12].
June 19-20, 2002	First national NEES consortium development project workshop [12].
September-October 2002	In award (NSF 01-164), NSF makes five awards to five universities for new equipment and upgrades totaling $15.5 million [12].
January 2003	NEES consortium is incorporated; initial directors and bylaws are chosen [12].
April 28, 2003	First NEES Consortium, Inc., election [12]
May 21-22, 2003	First annual meeting of NEES Consortium, Inc. [12]
September 30, 2004	Planned completion date for NEES construction.
FY 05–FY 14	NEES research and operations period.

the original name for NEES) was developed by NSF[9] with a project-development plan in July 1997. Later that summer, the NSB gave the go-ahead to the NSF Engineering Directorate (ENG) to develop a detailed project-management proposal.

In October 1997, another NEHRP Reauthorization Act (PL 105-47) was signed into law; it required NSF and other agencies to develop a "comprehensive plan for earthquake engineering research." An EERI meeting was held in San Francisco by Jim Jirsa (University of Texas, Austin) to assist NSF in planning for NHPS. At the meeting, William Anderson (director of the ENG Earthquake Mitigation Program) reported on the favorable review by the NSB. Over the next year, several workshops and meetings were held to discuss NHPS and engage the relevant research communities. In June 1998, Eugene Wong, an information-technology specialist formerly with the Office of Science and Technology Policy (OSTP), assumed the position of assistant director of ENG. Additional planning workshops for NHPS/NEES were also held by the earthquake engineering community in 1998 [2a, 9].

NEES received MREFC funding in the FY 2000 budget. NSF began issuing solicitations for proposals in December 1999 and made the first 11 equipment awards between September 2000 and February 2001. After a system integration scoping study was completed, the award for full system integration was made in August 2001. The award for consortium development was made in September 2001. Five additional equipment awards were made in September 2002. All NEES equipment will be fully operational by September 30, 2004.

References

[1] Assessment of Earthquake Engineering Research and Testing Capabilities in the United States. Earthquake Engineering Research Institute, Proceedings: Document WP-01A, Summary Report: Document WP-01. September 1995.

[2a] Geotechnical Earthquake Engineering Experimental Facilities: Developing a National Network with Structural, Seismological and Coastal High Performance Seismic Simulation Facilities, University of California, Davis, May 28-29, 1998. Available at <http://cgm.engr.ucdavis.edu/NEES/NHPSconference.html>.

[2b] Developing a National Network with Structural, Seismological, and Coastal Earthquake Engineering Seismic Simulation Facilities. University of California, Davis, April 1999 (workshop date May 1998).

---

[9]Bruce Kutter: "A National Network for High Performance Seismic Simulation (NHPS) was proposed within NSF as a Major Research Equipment (MRE) initiative to respond to the need to develop integrated experimental research facilities" [2b]. Bruce Kutter: "Within the National Science Foundation, the focus of NEES has been steered toward an ultimate goal of non-physical simulations" [2b]. In 1999, Rita R. Colwell indicated that NEES was modeled after the "highly successful" National Nanofabrication Users Network [3].

[3] Federal News Service. Prepared Testimony of Dr. Rita Colwell, director, National Science Foundation, before the Subcommittee on Veterans Affairs, Housing and Urban Development, and Independent Agencies, Senate Committee on Appropriations, March 23, 1999.

[4] Network for Earthquake Engineering Simulation (NEES) program briefing, Roundtable discussion by earthquake engineering researchers, NSF, February 10, 2000.

[5] SRI International Science and Technology, History of the NSF Earthquake Hazard Mitigation Program. Prepared for NSF, Draft Report, October 7, 1999, p. 43.

[6] EERI Newsletter. NSB Gives NSF's earthquake mitigation program green light to continue planning a national network for high performance seismic simulation, November 1997.

[7] J. Jirsa and D. Abrams. Planning meeting scheduled for June regarding national network of high performance seismic simulation facilities. EERI News, April 1998.

[8] An Experimental Facilities Initiative in Earthquake Engineering: Action Plan for Upgrading, Expansion and Utilization. Report to the NSF, January 1996.

[9] Report for a National Science Foundation Workshop for Tsunami Research Facilities. NSF Workshop Report, 1998.

[10] Federal News Service. Prepared Testimony of Dr. Joseph Bordogna, acting deputy director, National Science Foundation, before the Basic Research Subcommittee, House Committee on Science, Feb. 23, 1999. Prepared Testimony of Dr. Eugene Wong, assistant director for engineering, National Science Foundation, Subcommittee on Science, Technology and Space; Senate Committee on Commerce, Science and Transportation, June 29, 1999. (These two statements are essentially identical.)

[11] White Paper: Towards a Vision for the NEES Collaboratory, version 3.0, Task Group on Collaboratory Research, NEES Consortium Development Project, CUREE; Oct. 1, 2002.

[12] NSF's NEES website: <www.eng.nsf.gov/nees>. NEES Consortium, Inc., website: <www.nees.org>. NEESgrid website: <www.neesgrid.org>.

[13] Network for Earthquake Engineering Simulation (NEES): Earthquake Engineering Research Equipment, Program Solicitation (NSF 00-6) December 20, 1999.

[14] Network for Earthquake Engineering Simulation (NEES): System Integration, Program Solicitation (NSF 00-7) December 2, 1999.

[15] NEHRP Coalition in Support of Reauthorization of the National Earthquake Hazards Reduction Program. Testimony of Dr. Daniel P. Abrams, Professor of Civil Engineering at the University of Illinois at Urbana-Champaign, before House Committee on Science, Subcommittee on Basic Research, April 24, 1997.

[16] NSF Press Release (NSF PA 98-2), April 28, 1998.

NEON (NATIONAL ECOLOGICAL OBSERVATORY NETWORK)

Description

The National Ecological Observatory Network (NEON) will be a continental-scale research platform consisting of geographically distributed observatories that are networked via state-of-the-art communications. NEON will allow researchers to study the structure and dynamics of US ecosystems with the goal of measuring and forecasting biologic change resulting from human and natural influences on local to continental scales. The overall conceptual theme of this network of research observatories will be the nature and pace of biologic change.

NEON observatories will contain cutting-edge instrumentation, site-based experimental infrastructure, natural-history and data archives; and computational, analytic, and modeling facilities. In addition to field-based infrastructure, NEON will include laboratory equipment and support personnel, and it will stimulate the development of technologies to permit new ways of integrating, analyzing, and visualizing data. Rather than being envisioned as multiple collections of sensors that monitor individual sites or regions, NEON will be a network of colocated infrastructure deployed across the United States, creating a "network of networks." Once commissioned, NEON will be used to conduct research projects supported through disciplinary and multidisciplinary programs at NSF. Data generated from "standard measurements" made with NEON will be publicly available. NEON will transform ecologic research by enabling research on regional to continental scales using state-of-the-art technology.

Approval and Funding History

Funds for construction were requested in FY 2001 and FY 2003 budgets but were denied without prejudice by Congress. Negotiations are under way for the inclusion of funding in the FY 2004 budget.

Managing Institutions

Not applicable.

Development Summary

Beginning in 1997, as part of its annual strategic planning process, the NSF Biological Sciences Directorate (BIO) senior management discussed the infrastructure needed to enable leading-edge biologic research. Bruce Hayden, who had recently joined NSF as a visiting-scientist director of the Division of Environmental Biology, suggested that a national network of environmental observatories would enable ecologists to address important regional- and continental-scale questions. See Table C-10 for a timeline of the major developments.

He then informally discussed these ideas at the annual meeting of the Ecological Society of America (ESA). With favorable responses from the research community, BIO began to develop this idea into NEON. Scott Collins, a former Ecology Program director, succinctly summed up NEON's development [1]: "Once we had the concept . . . we asked the community to design it for us."

In March 1998, the President's Committee of Advisors on Science and Technology (PCAST) Panel on Biodiversity and Ecosystems, which

TABLE C-10 Timeline of Major Developments

August 1997	BIO senior management discuss infrastructure needed to enable leading-edge biologic research as part of annual strategic planning cycle; Bruce Hayden, director of Division of Environmental Biology, suggests that national network of colocated infrastructure is needed to address next frontier in ecologic research, regional- to continental-scale studies; he floats his idea at annual ESA meeting; BIO begins to formulate idea into NEON [2].
March 1998	PCAST Panel on Biodiversity and Ecosystems completes report *Teaming with Life,* many elements of which appear to have been incorporated into NEON; panel included Peter Raven (chair) and Rita R. Colwell [3].
April 1998	At April 6-7 BIO Advisory Committee meeting, Bruce Hayden provides report on senior management planning, including discussion of environmental observatories.
August 12, 1998	NSB establishes TFE under its Committee on Programs and Plans [4].
September 10-11, 1998	First BON workshop [5].
November 1998	NEON identified to NSB on list of potential large-infrastructure projects.
January 14, 1999	TFE public hearing in Portland, Oregon [4].
January 14-17, 1999	Second BON workshop [6].
February 17-18, 1999	Public NSB symposium in Los Angeles, California, provides community feedback for TFE [4].
March 8, 1999	TFE public town-hall meeting in Arlington, Virginia [4].
April 1999	At April 22-23 BIO Advisory Committee meeting, Mary Clutter discusses proposals for FY 2001 budget, including establishment of National Ecological Observatory Network (NEON).
May 6-7, 1999	Third BON workshop [7].
June 29, 1999	TFE's interim report is approved by NSB and released for public comment [4].
August 1999	NSB approves NSF FY 2001 budget request, including NEON as MREFC project.
August 27-29, 1999	Fourth BON workshop [8].
January 10-12, 2000	First NEON workshop, on basic concept development (26 participants, four NSF observers) [9].

TABLE C-10 Continued

February 2, 2000	NSB approves and releases TFE's final report, *Environmental Science and Engineering for the 21st Century* [4].
February 2000	NEON appears in NSF's FY 2001 budget, which is later denied "without prejudice" by Congress.
March 9-10, 2000	Second NEON workshop, on equipment, infrastructure, and personnel (24 participants) [10].
May 3-4, 2000	Third NEON workshop, on organization and management (13 participants) [11].
2001	National Research Council report: *Grand Challenges in Environmental Research.*
February 2002	NEON appears in NSF FY 2003 budget, which is later denied "without prejudice" by Congress.
June 4-5, 2002	Fourth NEON workshop, on standard measurements and infrastructure needs (22 participants, three observers) [12].
June 14-16, 2002	Fifth NEON workshop, on biologic-collections community (30 participants) [13].
August 5, 2002	NEON session at ESA, Tucson, Arizona: NEON: Next Steps toward Reality [18].
September 16-18, 2002	Sixth NEON workshop, on information management (19 participants, three observers) [14].
September 2002	NSF awards AIBS $1.3 million to create IBRCS with AIBS Executive Director Richard O'Grady as principal investigator [15].
September 10, 2002	ESA sends letter to Rep. Alan Mollohan (D-WV), ranking member, House Subcommittee on Veterans Affairs, Housing and Urban Development, and Independent Agencies, asking for his support for NEON [16].
November 15-16, 2002	IBRCS holds first face-to-face meeting in Arlington, Virginia [17].
December 13, 2002	IBRCS holds NEON town-hall meeting in Washington, D.C. [17].
January 17, 2003	IBRCS holds NEON town-hall meeting in Los Angeles, California [17].
February 14, 2003	IBRCS holds NEON town-hall meeting in Denver, Colorado [17].
February 2003	NEON appears in FY 2004 budget request.
March 25, 2003	IBRCS white paper released at public roundtable at National Press Club [15, 17, 18].

continued

TABLE C-10 Continued

May 2003	National Research Council Board on Life Sciences begins study on NEON, to be completed in fall [21].
June 30, 2003	Elizabeth Blood new NSF program director for research resources and is responsible for NEON [17].
July 2003	House Passes NSF bill funding $12.0M for NEON.
September 2003	Senate Subcommittee on Veterans Affairs, Housing and Urban Development, and Independent Agencies of the Committee on Appropriations recommends no new starts in MREFC account [19].
September 4-5, 2003	Seventh NEON workshop, on NEON coordination and implementation [18].
September 17, 2003	National Research Council Board on Life Sciences report on NEON released [20, 21].

included Peter Raven (chair) and Rita R. Colwell, completed the report *Teaming with Life*. Many elements of that report appear to have been incorporated into the NEON project. In August 1998, the NSB Committee on Programs and Plans established a Task Force on the Environment [4] (TFE), which included Clutter. After a series of public meetings and several opportunities for community input, the TFE published its final report, *Environmental Science and Engineering for the 21st Century*, in February 2000. Soon thereafter, NEON was described as a first step in fulfilling the vision outlined in the report. And in 2001, the National Research Council report *Grand Challenges in Environmental Sciences* not only called for regional and continental approaches for the eight grand challenges identified in the report but also suggested that infrastructure would be needed to enable such research.

While the TFE was in operation, a series of four workshops [5-8] were held for the development of the Biodiversity Observation Network (BON). In January 2000, the first NEON workshop was held [9], and BON was incorporated as a small part of the much broader NEON project. In February 2000, NEON was included in the NSF FY 2001 budget. Two more NEON workshops [10, 11] were held in the first half of 2000 before Congress denied funding "without prejudice." NEON was not included in the FY 2002 budget but was included in the FY 2003 and was again denied funding "without prejudice." Three more NEON workshops were held from July to September 2002 [12-14].

In September 2002, ESA sent a letter to Rep. Alan Mollohan (D-WV), ranking member of the House Subcommittee on Veterans Affairs, Housing and Urban Development, and Independent Agencies, asking him to support NEON. In that same month, NSF awarded the American Institute of Biological Sciences (AIBS) $1.3 million to establish a working group to design the Infrastructure for Biology at Regional to Continental Scales (IBRCS) with AIBS Executive Director Richard O'Grady as principal investigator. The goal of the IBRCS working group, chaired by Kent Holsinger (University of Connecticut), is to "further advance the [NEON] initiative by explaining the scientific rationale behind the need for NEON, how NEON will operate to meet that need, and the results that NEON is expected to produce" [17]. The IBRCS began holding a series of three town-hall meetings on NEON in December 2002 and released a white paper on NEON in March 2003 at a public roundtable.

In February 2003, NEON was included in the NSF FY 2004 budget. In June 2003, Elizabeth Blood became the new NSF program director for research resources, and she is responsible for NEON. In fall 2003, the National Research Council Board on Life Sciences released a study on NEON. The author committee strongly supported the creation of a NEON-like program and commended NSF's overall vision for NEON. It also cautioned that the proposed implementation plans needed modification and refinement to ensure that NEON would focus on the most important scientific issues, efficiently provide the national network of infrastructure essential for each challenge, encourage creative research, and meet the requirements for MREFC funding.

References

[1] Jeffrey A. Goldman. NEON Illuminated (editorial). *BioScience*, 53:447.

[2] J. Mervis and J. Kaiser. NSF hopes Congress will see the light on NEON. *Science* 300:1869.

[3] Teaming with Life: Investing in Science to Understand and Use America's Living Capital, Panel on Biodiversity and Ecosystems, President's Committee of Advisors on Science and Technology, March 1998.

[4] National Science Board, Environmental Science and Engineering for the 21st Century: The Role of the National Science Foundation, February 2, 2000: NSB 00-22.

[5] Final Report: Biodiversity Monitoring Project Workshop. A National Biodiversity Observatory Network. University of Virginia's Blandy Experimental Farm, September 1998.

[6] Report of the Second Workshop on the Biological Observation Network. National Center for Ecological Analysis and Synthesis, Santa Barbara, Calif., January 1999.

[7] Report of the Third Workshop on the Biodiversity Observatory Network. California Academy of Sciences, San Francisco, Calif., May 1999.

[8] Report of the Fourth Workshop on the Biological Observation Network, National Center for Ecological Analysis and Synthesis, Santa Barbara, Calif., August 1999.

[9] Report on the First Workshop on the National Ecological Observatory Network (NEON), January 10-12, 2000. Held at Archibold Station, Lake Placid, Fla. Sponsored by the National Science Foundation.

[10] Report to the National Science Foundation from the Second Workshop on the Development of a National Ecological Observatory Network (NEON), March 9-13, 2000. Held at San Diego Supercomputer Center, La Jolla, Calif.

[11] Report to the National Science Foundation from the Third Workshop on the Development of a National Ecological Observatory Network (NEON), May 3-4, 2000. Held at Santa Fe Institute, Santa Fe, N. Mex.

[12] Report to the National Science Foundation from the Fourth Workshop on the Development of a National Ecological Observatory Network (NEON): Standard Measurements and Infrastructure Needs, June 4-5, 2002. Held at The Millennium Hotel, Boulder, Colo.

[13] Final Report NEON-V: CRIPTON Workshop. Collections, Research, Inventories, and People for Taxonomic Opportunities in NEON, June 14-16, 2002. Held at the Field Museum of Natural History, Chicago, Ill.

[14] Report to the National Science Foundation from the Sixth Workshop on the Development of a National Ecological Observatory Network (NEON): Information Management, September 16-18, 2002. Held at National Center for Ecological Analysis and Synthesis, University of California, Santa Barbara.

[15] Sonya Senkowsky. NEON: Planning for a New Frontier in Biology, *BioScience* 53:456.

[16] Ann M. Bartuska. ESA Statements, September 10, 2002. Available at <http://www.esa.org/pao/statements_resolutions/statements/nsfneon.htm>.

[17] K.E. Holsinger and the IBRCS Working Group. IBRCS White Paper: Rationale, Blueprint, and Expectations for the National Ecological Observatory Network, Washington, D.C.: American Institute of Biological Sciences, March 2003.

[18] IBRCS web site: <http://ibrcs.aibs.org>.

[19] Senate Report 108-143, September 2003.

[20] Panel Suggests a Difference Shade of NEON. *Science* 301:1828.

[21] National Research Council. NEON: Addressing the Nation's Environmental Challenges. Washington, D.C.: The National Academies Press, 2003. Available at <http://books.nap.edu/catalog/10807.html>.

OOI (OCEAN OBSERVATORIES INITIATIVE)

Description

The Ocean Observatories Initiative (OOI) will "provide the ocean science community in the U.S. with the basic infrastructure required to make long-term measurements in the oceans" [1]. The OOI will consist of three components: a global network of relocatable deep-sea observatories based around a system of moored buoys, a system of cabled permanent observation sites on the seafloor spanning regional-scale (10-1000 km) features, and an expanded network of coastal observatories. Scientific questions in ocean research and a growing awareness of the interconnectedness of the ocean and land environments essential for sustaining the human race have prompted an increased desire for long-term, temporal information about ocean systems. Driven primarily by the 2001 NSF Division of Ocean

Sciences (OCE) Decadal Committee report *Ocean Sciences at the New Millennium*, the OOI addresses needs identified by the ocean-science community for advancing basic ocean research. In the future, the OOI will become the research-oriented contribution to the proposed Integrated and Sustained Ocean Observing System (IOOS) developed by Ocean.US under the auspices of the National Oceanographic Partnership Program (NOPP). The IOOS will serve as the key US contribution to the international Global Ocean Observing System [1]. The OOI will develop in parallel with the Ocean Information Technology Infrastructure program, which will support its data assimilation, archiving, analysis, and visualization needs [2].

Approval and Funding History

Support for OOI design and development began in 2001, and the OOI was identified as a FY 2006 new start in the NSF 2004 budget request.

Managing Institutions

Not applicable.

Development Summary

The OOI, which has yet to receive NSF MREFC funding, represents the consolidation of various recommendations made by workshops and reviews. In the early part of the 1990s, NSF-funded workshops broadly discussed the future need for observatories in the geophysical sciences. In 1996, a subset of those attending prior conferences met to discuss the possibility of a national seafloor observatory system. See Table C-11 for a timeline of the major developments. That led to the formation of the Deep Earth Observatories on the Seafloor (DEOS) initiative in 1997. The initial focus of DEOS limited itself to deepwater geo-observatories, but it gradually expanded to include nearshore observatories and water-column studies. In 1999, the name was changed to Dynamics of Earth and Ocean Systems to reflect the "effort to engage the wider oceanographic community" [3].

In 1999, NSF asked the National Research Council to investigate "the scientific merit, technical requirements, and overall feasibility" [3] of developing an unmanned seafloor observatory. Drawing on internal resources, past reports, and recommendations from the January 2000 Symposium on Seafloor Observatories, the Research Council issued the 2000 report *Illuminating the Hidden Planet*. The report provided 10 recommendations for moving forward with a seafloor observatory program to collect time-series observations using both moored-buoy and cabled observato-

TABLE C-11 Timeline of Major Developments

Before 1996	Discussions among geophysical scientists concerning future observatory needs.
1996	Meeting to discuss possibility of seafloor observatory [3].
1997	Establishment of Deep Earth Observatories on the Seafloor (DEOS) initiative [3].
1997	Congress establishes NOPP [4].
1998	Formation of NSF OCE Decadal Committee [5].
1999	DEOS changed to stand for Dynamics of Earth and Ocean Systems [3]; NSF asks National Research Council to investigate possibility of unmanned seafloor observatory [3].
January 10-12, 2000	Symposium on Seafloor Observatories in Islamorada, Florida.
2000	National Research Council releases *Illuminating the Hidden Planet* [3].
January 8-9, 2001	Ocean Observatories Steering Committee meeting in Washington, D.C. [6].
March 2001	NSF OCE Decadal Committee releases *Ocean Sciences at the New Millennium* [5].
June 18-19, 2001	Ocean Observatories Steering Committee meeting in Boulder, Colorado [7].
July 12, 2001	Hearing on ocean exploration and ocean observations before House Committees on Resources and Science [4].
March 19, 2002	DEOS Steering Committee meeting with NSF Director Colwell [8].
Spring 2002	NSF requests National Research Council study on implementing seafloor observatory network [1].
April 16-17, 2002	DEOS Steering Committee meeting in La Jolla, California [8].
May 7-9, 2002	CoOP meeting in Savannah, Georgia [9].
August 26-28, 2002	SCOTS meeting in Portsmouth, Virginia [10].
October 3-4, 2002	DEOS Steering Committee meeting in Washington, D.C. [9].
February 5-7, 2003	Moored Buoy Working Group Meeting in Santa Fe, New Mexico [11].
February 2003	President Bush's FY 2004 budget request includes out-year funding for OOI in FY 2006.
March 3-4, 2003	DEOS Steering Committee meeting in Washington, D.C. [11].
June 3-4, 2003	DEOS Steering Committee meeting in Washington, D.C. [11].
June 19-20, 2003	Cable Re-Use Committee meeting in Washington, D.C. [11].
Summer 2003	National Research Council releases *Enabling Ocean Research in the 21st Century*.

ries and having the open availability of data and coordination with international efforts.

The NSF OCE Decadal Committee convened in 1998 released its recommendations for the future of ocean research in March 2001. The "New Technologies" section cited three elements of technologic development that would later make up the three components of the OOI. Recognizing that "ships alone do not represent scientific capability" and that "physical and chemical sensors for making in situ measurements" would prove crucial to the advancement of ocean sciences, the report listed the need for the following capabilities: the ability to conduct ocean sampling "on space and time scales appropriately tuned to the process being investigated, new fixed (cabled or moored buoy) and mobile (ROV [Remotely operated vehicle], AUV [Autonomous underwater vehicles], and drifter) ocean-observing systems, [and] long-term monitoring of the ocean" [5]. In a Hearing before the House Committees on Resources and Science on July 12, 2001, Rita R. Colwell, director of NSF, cited the decadal study when introducing the recently established OOI. She described the effort as a means to "provide basic infrastructure for a new way of gaining access to the oceans, by starting to build a network of ocean observatories that would facilitate the collection of long time-series data streams needed to understand the dynamics of biological, chemical, geological and physical processes" [4].

Before the July hearing, the Ocean Observatories Steering Committee (OOSC) had held two meetings, in January and June 2001, to clarify its role in the OOI development process. The OOSC clarified its position as the point of contact between NSF and the community of ocean observatories [9]. The June meeting also established that with respect to NSF "the purpose of the OOSC is to advise the NSF on the MRE initiative to fund infrastructure for ocean observatories" [7].

The DEOS Steering Committee met with NSF Director Colwell on March 19, 2002. Dr. Colwell recommended that DEOS not seek earmarked funding for the OOI inasmuch as it would "not include the additional funding through the R&RA account that typically accompanies MREFC projects" [8]. The budget request for FY 2003 submitted by the president in February 2002 did not include funding for the OOI from the NSF MREFC account.

In the spring of 2002, NSF requested a study from the National Research Council on "issues related to the implementation of a seafloor observatory network" [1]. The particular concerns addressed included the development and implementation of the network, the impact on existing ocean-studies facilities, and the potential role of the OOI in IOOS and other international efforts. A preliminary version of the report was made available in summer 2003 and was titled *Enabling Ocean Research in the 21st*

Century. The recommendations made by the Research Council draw on three workshops overseen by the DEOS Steering Committee.

In May 2002, a Coastal Ocean Processes (CoOP) workshop took place in Savannah, Georgia. The purposes of the workshop included determining which science is best suited to coastal observing systems and identifying technologies most useful to the Coastal Integrated Observing System. The design criteria suggested included a set of relocatable, or "Pioneer," arrays that the CoOP considered implementing as part of the OOI coastal component [9].

In August 2002, a Scientific Cabled Observatories for Time Series (SCOTS) workshop took place in Portsmouth, Virginia. The workshop's charge included providing "advice on the scientific rationale and implementation of a network of regional cabled observatories" [10].

In February 2003, a Moored Buoy Working Group met in Santa Fe, New Mexico. During the same month, the President submitted his request for the FY 2004 budget, which did not include MREFC funding for the OOI but did for the first time make out-year requests, including funding for the OOI in FY 2006.

References

[1] National Research Council. Enabling Ocean Research in the 21st Century, Washington, D.C.: The National Academies Press, p. 1, 2003.

[2] Ocean Observatories Initiative brochure.

[3] NRC. Illuminating the Hidden Planet. Washington, D.C.: National Academy Press, 2000.

[4] Hearing on Ocean Exploration and Observations. Testimony of Dr. Rita R. Colwell, director NSF, before House committees on Resources and Science, July 12, 2001.

[5] NSF OCE Ocean Sciences at the New Millennium, 2001.

[6] Minutes, OOSC, January 8-9, 2001, Washington, D.C.

[7] Minutes, OOSC, June 18-19, 2001, Boulder, Colo.

[8] Minutes, DEOS-SC April 16-17, 2002, La Jolla, Calif.

[9] Minutes, DEOS-SC October 3-4, 2002, Washington, D.C.

[10] SCOTS workshop report, August 26-28, 2002, Portsmouth, Va.

[11] DEOS website: <www.coreocean.org/DEOS>.

POLAR AIRCRAFT

Description

The Polar Aircraft project was required in order to modify three NSF-owned ski-equipped LC-130 aircraft to meet Air Force safety and operability standards. These modifications include engineering, avionics, airframe, safety, and propulsion. Ski-equipped LC-130 aircraft are the backbone of air transport for the US Antarctic Program (USAP) and support NSF research in the Arctic. The LC-130 is the only heavy-lift aircraft capable of making winter landings at the South Pole [1]. All polar LC-130 aircraft are operated by the New York Air National Guard (ANG) 109th Airlift Wing. ANG owns six LC-130s and also operates the four NSF-owned aircraft.

Approval and Funding History

The aircraft were approved in August 1998. MREFC funding began in FY 1999 and was continued in FY 2000 for planning, design, and development and construction. In FY 2001, NSF received authority to reprogram up to $1 million to complete the project.

Managing Institutions

The contract for the modifications is administered by the Air Logistics Command at Robins Air Force Base. The initial contractor was Raytheon Aircraft Integration Systems, which was later purchased by L3 Communications. L3 and about 240 subcontractors provide supplies and technical services.

Development Summary

In January 1997, a memorandum of agreement between the Navy, Air Force, National Guard Bureau, NSF, and Department of Defense identified ANG as the appropriate organization to assume operational control of all LC-130s in the USAP. See Table C-12 for a timeline of major developments. In March 1999, ANG assumed operational control for all LC-130s [3]. Six of the 10 aircraft are ANG-owned. One NSF-owned aircraft already met Air Force safety and operability standards, but three older NSF-owned aircraft needed additional upgrades to meet the standards.

During 1998, the NSF Office of Polar Programs reviewed whether the polar mission could be supported with nine rather than 10 LC-130s, but the review made clear that ANG required 10 LC-130s to support the polar mission in addition to its other missions [4]. Because that review was in progress when the FY 1999 NSF budget request was being prepared, the FY 1999 NSF budget request included $20 million in the Major Research Equipment account for the modification of two LC-130s, and these funds were appropriated in FY 1999. Funding for the third aircraft was provided in FY 2000 [5].

TABLE C-12 Timeline of Major Developments

January 21, 1997	Memorandum of agreement transfers control of LC-130 polar heavy airlift to ANG [2].
March 26, 1998	ANG assumes control of all LC-130 polar operations [3].
Summer 1998	Discussions regarding need to update all three NSF-owned aircraft [4].
1999	First year of MREFC funding to upgrade NSF-owned LC-130s to meet Air Force safety and operability standards.

References

[1] Global Security web site: <www.globalsecurity.org/military/systems/aircraft/lc-130.htm>.
[2] DefenseLINK News. Memorandum No. 007-M, January 21, 1997.
[3] DefenseLINK News. News Release No. 132-98, March 26, 1998.
[4] Preliminary Report of the August 12-13, 1998, NSB Meeting (NSB 98-164).
[5] NSF FY 2001 MRE Budget Request.

POLAR CAP OBSERVATORY

Description

The Polar Cap Observatory, a multi-instrumented ground-based facility, will be in the northern polar cap at Resolute Bay in Canada. It will consist of a large state-of-the art radar facility with an accompanying array of smaller optical and radiowave remote-sensing instruments. The new facility will allow for monitoring of "space weather," the conditions in the space environment that can influence the performance and reliability of spaceborne and ground-based technologic systems. Space-weather storms can disrupt satellites, communication, navigation, and electric-power distribution grids.

Approval and Funding History

The project was approved in May 1998 and placed in FY 1998 and FY 1999 budget requests, but no funding was received. The project has not been included in a budget request since then.

Managing Institutions

Not applicable.

RSVP (RARE SYMMETRY VIOLATING PROCESSES)

Description

The Rare Symmetry Violating Processes project (RSVP) will consist of two experiments at the Brookhaven National Laboratory (BNL) Alternating Gradient Synchrotron (AGS) that will look for rare decay processes of certain elementary particles. If observed, these processes would indicate the existence of new phenomena beyond the Standard Model (SM) of elementary particle physics. Probing rare processes generally involves effects due to virtual-particle production and annihilation, and through these effects one is provided access to particle mass scales much higher than those accessible through direct accelerator production. Initially, the RSVP would construct and carry out two fundamental experiments: MECO (a study of an extremely rare process for conversion of muons to electrons, hence the acronym) and KOPIO (a similarly rare process for studying the decay of neutral kaons into neutral pions). MECO will search for conversion of muons into electrons in the nuclear Coulomb field, an event with a 10^{-17} probability of occurring. Muon-to-electron conversion is accommodated within the SM, but the basic SM mechanism would produce an event rate far below what is measurable. Instead, MECO will search for excess conversion that would point toward new physics, that is, beyond the SM. KOPIO will explore the world of Charge-Parity (CP) violation, the process by which the observed matter-antimatter asymmetry is thought to have arisen. KOPIO will search for rare decays of neutral kaons into neutral pions and neutrino-antineutrino pairs, a process mediated by direct CP violation and very well understood in terms of the SM. Any deviation from the SM or from similar measurements of CP violation in the B-meson sector would indicate the existence of physics beyond the SM. By either ruling out or characterizing these processes at probability as low as 10^{-17}, the RSVP will help to clarify existing questions and elicit new ones about the fundamental structure of matter. The RSVP will represent the efforts of a 30-institution collaboration involving the United States, Canada, Switzerland, Italy, Japan, and Russia.

Approval and Funding History

Although recommended for approval by the NSB in 2001, overall national budget pressures have delayed budget appropriations for the RSVP. Funding is anticipated in FY 2006, as was indicated in the FY 2004 budget request to Congress.

Managing Institutions

RSVP will be an MREFC project carried out under a cooperative agreement between NSF and New York University (NYU), the RSVP grant-holding institution. A memorandum of understanding and subcontract between NYU and the University of California, Irvine (UCI) exists to oversee the construction of MECO. UCI will be the lead institution for MECO. A memorandum of understanding and subcontract between NYU and the State University of New York at Stony Brook (SUNY-SB) exists for the construction of KOPIO. SUNY-SB will be the lead institution for KOPIO. As the site for both experiments, BNL will assume a support and oversight role in RSVP.

Development History

RSVP will bring together two experiments that seek to detect rare processes that violate symmetries required by the SM of particle physics. The history of MECO can be traced to a 1989 idea that led to a 1992 design proposal for implementation in the Moscow Meson Factory (MMF) [1] [2]. Because of changes in government, the project did not come to fruition. See Table C-13 for a timeline of the major developments. A 1997 paper presented at the Stanford Linear Accelerator Center (SLAC) Summer School discussed the proposal to implement MECO at BNL AGS [1]. The KOPIO experiment was developed as a means of improving understanding of the observed preference in the universe for matter over antimatter [3].

In 1999, a joint MECO-KOPIO proposal was submitted as a single proposal called RSVP through NYU, with John Sculli as principal investigator, for consideration by NSF. In May 2000, the MREFC panel of the NSF Directorate for Mathematical and Physical Sciences recommended to the NSF director that the request for funding for RSVP be included in the FY 2002 budget request to Congress. In October 2000, the NSB approved RSVP as a candidate to be included as an MREFC project in the NSF budget in FY 2002 and beyond.

In late January 2002, the High Energy Physics Advisory Panel (HEPAP) to the Department of Energy and NSF endorsed the scientific goals of RSVP in its 20-year roadmap for the field. RSVP was not included in the MREFC FY 2002 budget request (also January 2002), because of budget constraints.

Starting in FY 2001, NSF has funded R&D activities for the RSVP through merit-reviewed R&D proposals. The funding profile was about $900,000 per year for each of FY 2001, 2002, and 2003. Additional funds have been requested and are under review. NSF has held periodic reviews

TABLE C-13 Timeline of Major Developments

1989	Russian scientists propose theory for muon-to-electron conversion experiment [1].
1992	Proposal for Russian muon-to-electron experiment at MMF [2].
Summer 1997	SLAC Summer School presentation of MECO proposal [1].
October 1999	RSVP proposal submitted for MREFC funding [4, 5] through NYU (J. Sculli, principal investigator).
November 1999 – November 2000	NSF conducts three reviews of RSVP proposal covering scientific merit, technical issues, and management; no action taken.
May 2001	MECO R&D proposal funded at $0.5 million/year for 3 years (FY 2001-2003) to UCI (W. Molzon, principal investigator).
June 6, 2001	House Subcommittee on Research hearing during which MREFC board's approval of RSVP is affirmed [6].
July 2001	KOPIO R&D proposal funded at $0.4 million/year for 3 years (FY 2001-2003) to Yale University (M. Zeller, principal investigator).
January 2002	HEPAP endorses scientific goals of RSVP in its long-range planning (roadmap) for US high-energy physics (*The Science Ahead—The Way to Discovery*).
September 2002	KOPIO design and development proposal funded at $0.3 million/year for 2 years (FY 2002-2003) to SUNY/SB (M. Marx, principal investigator).
October 2002	MECO design and development proposal funded at $0.3 million/year for 2 years (FY 2002-2003) to UCI (W. Molzon, principal investigator).
January 2003	NSF budget proposal for FY 2004 makes out-year request for RSVP MREFC funding in FY 2006.
September 2003	MECO design and development proposal funded at $0.5 million for 1 year to UCI (W. Molzon, prinicipal investigator); KOPIO design and development proposal funded at $0.5 million for 1 year to SUNY/SB (M. Marx, principal investigator).

of technical developments, the RSVP management plan, and R&D activities in an effort to ensure readiness for MREFC funding when it becomes available. Several additional university groups have joined the RSVP collaboration. RSVP is not included in the FY 2004 budget request, but it does appear as an out-year approval for FY 2006 funding. Design and development continue.

References

[1] William Molzon. Improved Test of Muon and Electron Number Conservation in Muon Processes. Proceedings of the 1997 SLAC Summer School Topical Conference, 1997.
[2] John Sculli. μ -> e Conversion Status and Prospects. Proceedings from the Workshop on Physics at the First Muon Collider and at the Front End of the Muon Collider, 1998.
[3] The KOPIO Experiment, RSVP website. Available at <www.bnl.gov/rsvp/KOPIO.htm>.
[4] Approved minutes from meeting of the Advisory Committee of the NSF Directorate for Mathematical and Physical Sciences, April 12-13, 2001.
[5] Approved minutes from meeting of the Advisory Committee of the NSF Directorate for Mathematical and Physical Sciences, November 1, 2001.
[6] Hearing Summary from the NSF OLPS of the House Subcommittee on Research hearing on the NSF Research and Related Activities Account and Plant Genomics, June 6, 2001.
[7] American Institute of Physics FYI 60(May 15, 2002).
[8] American Physical Society News Online, July 2002. Available at <www.aps.org/apsnews/0702/index.html>.

SPSE (SOUTH POLE SAFETY AND ENVIRONMENTAL PROJECT)

Description

The South Pole Safety and Environmental project (SPSE) addressed urgent safety concerns at the Amundsen-Scott South Pole Station. The project included replacement of the heavy-equipment maintenance facility, the power plant, and fuel-storage facilities.

Approval and Funding History

MREFC funding of $25 million was provided in FY 1997, and an additional $500,000 was provided in FY 2002 to complete the project.

The SPSE received $25 million for FY 1997 [6] to undertake emergency upgrades, including a new garage and shop, new fuel-storage tanks, and a new power plant [7]. Construction for the SPSE began in the Antarctic in the summer of 1998 and proceeded on schedule despite the bitter conditions of the polar environment. The final phase of the project, completion of the new power plant, ended in January 2001 [8].

SPSM (SOUTH POLE STATION MODERNIZATION)

Description

The South Pole Station Modernization project (SPSM) is a new research station to replace aging facilities at the South Pole. An elevated station will replace the 1975 dome that now houses the US-operated South Pole research facility. Built using a modular design, the new station will house

150 people, about 50 percent of whom will be scientists. Perhaps the most remarkable feature of the new station, however, will be the ability to raise the entire structure. Because of the snow that blows continuously across the flat Antarctic plain, all buildings eventually find themselves buried. Perching atop stilts, the new structure will stand above the drifting snow, slowing the buildup process. If the snow rises, the station can be raised even higher. The new facility is intended to accommodate the US Antarctic Program (USAP) for 25-40 years [1].

Approval and Funding History

MREFC funding for construction began in FY 1998.

Managing Institutions

The Amundsen-Scott South Pole Station is part of the USAP managed by NSF.

Development Summary
(includes South Pole Safety and Environmental Project, SPSE)

The scope of the USAP has increased dramatically over the 40 years of US presence at the southernmost point on the globe [2]. The Amundsen-Scott South Pole Station, originally intended to house a summer population of about 30 people, now accommodates over 200 in the summer and up to 50 in the winter [1]. See Table C-14 for a timeline of major developments.

In 1995, citing budgetary constraints, the Senate Subcommittee on Veterans Affairs, Housing and Urban Development, and Independent Agencies of the Committee on Appropriations requested a study from the National Science and Technology Council (NSTC) to review US Antarctic policy. In particular, the charge requested that the NSTC examine increasing international cooperation, reducing the year-round operations, and closing one or more of the South Pole stations.[10] The study, released in April 1996, recommended a continued year-round US presence at the South Pole to preserve regional stability and enhance US foreign policy. The report also indicated, however, the need for establishing a realistic budget and management plan, and it recommended the creation of an external NSF panel to examine the future of the USAP [3]. After a July 1996 hearing on the USAP at the House Subcommittee on Basic Science of

---

[10]USAP facilities include Amundsen-Scott South Pole Station, McMurdo Station, Palmer Station, and two research vessels [3].

TABLE C-14 Timeline of Major Developments (includes SPSE)

1975	Opening of Amundsen-Scott South Pole Station [10].
September 1995	Senate Subcommittee on Veterans Affairs, Housing and Urban Development, and Independent Agencies of Committee on Appropriations requests report from NSTC to review US Antarctic policy [3].
April 1996	Release of NSTC report *United States Antarctic Program* [3].
July 23, 1996	Hearing on NSTC report before Subcommittee on Basic Science of House Committee on Science [4].
August 1996	External panel on USAP is convened [5].
1997	SPSE receives $25 million from MREFC funds [6].
March 12, 1997	House Committee on Science hearing on external panel report [5].
April 1997	Release of final external panel report *The United States in Antarctica* [2].
1998	SPSM receives $24.9 million from MREFC funds.
November 1998	SPSE construction begins [8].
June 9, 1999	NSF Office of Polar Programs director testifies before Subcommittee on Basic Research of House Committee on Science on USAP [7].
Antarctic summer 1999	Completion of SPSE fuel-storage project and shop [8].
November 2000	SPSM construction begins [8].
January 18, 2001	Completion of new satellite communication link [8].
January 20, 2001	Completion of SPSE power plant [8].
February 2003	First winter occupancy of Wings A-1 and A-2 of new station.
February 2004	Estimated occupancy of medical facility and computer laboratory in new station.

the Committee on Science [4], Congress requested the recommended examination. An external panel was convened in August 1996 [5].

The findings of the external panel concurred with the NSTC report on the need for continued year-round presence in Antarctica and for maintaining all three permanent US facilities. Its report encouraged international cooperation, but it stipulated that the United States should continue to build and manage the permanent facilities. To sustain the US presence, the report recommended a plan for building a new optimized station at the South Pole, to be completed by 2005. In response to those reports, NSF requested appropriations to fund the SPSM. The station was designed to house 100 persons but with an infrastructure capable of supporting 150.

MREFC funding for the SPSM began in FY 1998 with an appropriation of $24.9 million [9]. The NSB approved expansion of the 110-person station concept to 150 in 2002. Construction of the tower linking the elevated structure to the new SPSE facilities began in FY 2000 [8]. Adverse weather conditions have slowed the delivery of construction materials and resulted in a shift in estimated completion from 2005 to 2007.

References

[1] Josh Landis. To build a better station. *The Antarctic Sun*. December 19, 1999.
[2] Report of the USAP External Panel. *The United States in Antarctica*, April 1997.
[3] Report of the Committee on Fundamental Science, NTSC, *United States Antarctic Program*, April 1996.
[4] NSF OLPA hearing summary, July 23, 1996.
[5] U.S. Antarctic Program, 1996-1997. *Antarctic Journal of the United States Review 1997*.
[6] NSF MRE FY 2000 Budget Request.
[7] Testimony of Dr. Karl Erb, director of NSF OPP, before House Committee on Science, Subcommittee on Basic Research, June 9, 1999.
[8] NSF Press Release (NSF PR 01-04), January 24, 2001.
[9] SPSM Funding Profile. Available at <www.nsf.gov>.
[10] J. Rand, et al. Rebuilding the South Pole Station. *Civil Engineering Magazine Abstracts*, December 2000.

TERASCALE COMPUTING PROJECTS

Description

In FY 2000, NSF funded a Terascale Computing System (TCS) [1], the first NSF terascale system to be deployed by the NSF terascale computing systems activity. Based at the Pittsburgh Supercomputing Center (PSC), the TCS has a peak performance of 6 teraflops. When it was dedicated in October 2001, it was the second-most powerful computer in the world and the fastest one available for civilian research. The TCS employs 3,000 Compaq Alpha processors organized into 750 four-processor nodes. Aside from providing unprecedented computational speed, the TCS features 3.0 terabytes of total memory, 40 terabytes of primary storage, and 300 terabytes of disk and tape storage.

In FY 2001, NSF funded the Distributed Terascale Facility (DTF) [2], a geographically distributed, grid-enabled terascale computing system developed at four institutions: the National Center for Supercomputing Applications (NCSA), the San Diego Supercomputer Center (SDSC), Argonne

National Laboratory (ANL), and the California Institute of Technology (Caltech).

In FY 2002, the terascale computing systems activity funded creation of the Extensible Terascale Facility (ETF), providing for the integration and upgrade of the TCS and DTF resources in an extensible architectural framework. In FY 2003, three new awards added scientific instruments, large datasets, and additional computing power and storage capacity to the ETF, enhancing the scientific utility of the system. By the end of FY 2004, the ETF will include more than 20 teraflops of computing power distributed among nine sites, facilities capable of managing and storing approximately a petabyte of data, high-resolution visualization environments, advanced scientific instrumentation, and toolkits for grid computing.

Approval and Funding History

NSF received its first MREFC appropriation for the construction of terascale computing projects in FY 2000. The TCS was funded in FY 2000, and the DTF in FY 2001. MREFC funds in FY 2002 provided upgrades to the TCS and DTF facilities and created an extensible terascale system. MREFC funds in FY 2003 provided through the Terascale Extensions Program connected four additional sites to the ETF: Oak Ridge National Laboratory (ORNL), Texas Advanced Computing Center (TACC) at the University of Texas at Austin, Indiana University, and Purdue University.

Managing Institutions

The TCS was built by the PSC in partnership with Compaq. The DTF was built by the NCSA and the SDSC, with ANL and Caltech in partnership with IBM, Intel, Myricom, Qwest, Oracle, and Sun. The ETF is the integration of the TCS and DTF facilities and partners with the option to include new resource partners. New ETF partner sites added in FY 2003 include ORNL, University of Texas at Austin, Indiana University, and Purdue University.

Development Summary

In 1998, the Division of Advanced Computational Infrastructure and Research (ACIR), part of the NSF Directorate for Computer and Information Science and Engineering (CISE), held three workshops addressing issues related to terascale and petascale computing. See Table C-15 for a timeline of major developments. The meetings culminated in the report *Terascale and Petascale Computing: Digital Reality in the New Millenium*. A joint NSF-Department of Energy (DoE) workshop at the National

TABLE C-15 Timeline of Major Developments [18]

May 27, 1998	CISE ACIR Terascale Science Workshop in Arlington, Virginia.
July 9-10, 1998	CISE ACIR Terascale Workshop on Algorithms for the New Millennium in Arlington, Virginia.
July 15-16, 1998	CISE ACIR Terascale HPCC Software for the Next Millennium Workshop in Arlington, Virginia.
July 30-31, 1998	Terascale NSF and DOE workshop at National Academies [7].
February 24, 1999	PITAC report is released [8].
June 1999	NSTC IT^2 Working Group Implementation Plan is proposed in President's FY 2000 budget [9].
December 29, 1999	NSF program solicitation for TCS is issued [10].
August 3, 2000	Award for TCS granted to PSC in partnership with Compaq [1].
October 2000	Prototype system arrives at PSC [11].
January 18, 2001	NSF program solicitation for DTF is issued [13].
April 2001	TCS Prototype begins allocated use.
August 9, 2001	Award for DTF is granted to TeraGrid consortium of NCSA, SDSC, ANL, and Caltech [2].
October 29, 2001	TCS is dedicated and begins "friendly-user" period [12].
April 25, 2002	NSF sends letter requesting proposals for ETF [14].
April 2002	TCS begins allocated use.
October 10, 2002	NSB approves ETF award to NCSA, SDSC, ANL, Caltech, and PSC [15].
March 11, 2003	NSF issues terascale extensions solicitation [16].
September 29, 2003	NSF awards three terascale extensions to four sites [3].

Academies followed the NSF workshops and identified six components necessary for a high-performance computing environment, including scalable storage and data management and networking [7].

In February 1999, the President's Information Technology Advisory Committee (PITAC) issued a report emphasizing the need for high-performance computing systems to ensure continued US leadership in basic research [8]. Shortly thereafter, the National Science and Technology Council (NSTC) Information Technology for the Twenty First Century (IT^2) Working Group developed an implementation plan and timeline for the future of US computing capabilities to be proposed in the President's FY 2000 budget proposal. Among the deliverables listed in the projected timeline was a combined computing power of 10 teraflops by FY 2001 [9].

In 2000, NSF issued a solicitation for proposals to construct the TCS [10]. An award was made to PSC on August 3, 2000 [1], to construct TCS in partnership with Compaq. In October 2000, a 256-processor prototype system was installed, and it was later made available for allocated use during FY 2001 [11]. The full 3,000-processor system, named LeMieux, was dedicated on October 29, 2001 [12], and began full allocated use in April 2002.

In 2001, NSF issued a second program solicitation to construct a DTF [13]. This competition resulted in an award to NCSA, SDSC, ANL, and the Center for Advanced Computing Research (CACR) at Caltech. The initial DTF, named TeraGrid, included computers capable of 11.6 teraflops, disk-storage systems with capacities of more than 450 terabytes of data, visualization systems, and data collections—all integrated via grid middleware and linked through a high-speed optical network.

NSF entered the next stage of its terascale computing activity in 2002 by making an ETF award to expand the capabilities of the initial DTF sites and to integrate PSC's LeMieux system [15].

In 2003, NSF made an additional three awards to build on the ETF's capabilities [17]. The new awards fund the high-speed networking connections needed to share resources at Indiana University, Purdue University, ORNL, and TACC across the ETF infrastructure. Through the new awards, the ETF will put neutron-scattering instruments, large data collections and other unique resources, and additional computing resources within reach of the nation's research and education community.

References

[1] NSF Press Release (NSF PR00-53), August 3, 2000.

[2] NSF Press Release (NSF PR01-67), August 9, 2001.

[3] NSF Press Release (NSF PR03-107), September 29, 2003.

[4] NSF Blue Ribbon Panel report: From Desktop to Teraflop: Exploiting the US Lead in High Performance Computing, 1993. Quoted in CISE ACIR workshops summary report *Terascale and Petascale Computing: Digital Reality in the New Millenium*, 1998.

[5] Report of the Task Force on the Future of NSF Supercomputing Centers Program, 1995. Quoted in CISE ACIR workshops summary report *Terascale and Petascale Computing: Digital Reality in the New Millenium*, 1998.

[6] NSF Press Release (NSF PR 97-27), March 28, 1997.

[7] Department of Energy/National Science Foundation "National Workshop on Advanced Scientific Computation," National Academy of Sciences, Washington, D.C. July 30-31, 1998.

[8] PITAC report. Information Technology Research: Investing in Our Future. February 24, 1999.

[9] NSTC IT² Working Group implementation plan. *Information Technology for the Twenty-First Century: A Bold Investment in America's Future*, June 1999.

[10] Terascale Computing System, Program Solicitation (NSF 00-29) December 29, 1999.

[11] PSC News Release, January 29, 2001.

[12] PSC News Release, October 29, 2001.

[13] Distributed Terascale Facility, Program Solicitation (NSF 01-51), January 18, 2001.

[14] NSF Dear Colleague Letter on ETF (NSF 02-119), April 25, 2002.

[15] NCSA Press Release, October 10, 2002. SDSC Press Release, October 10, 2002.

[16] Terascale Extensions: Enhancements to the Extensible Terascale Facility (NSF 03-553), March 11, 2003.

[17] NSF Press Release (NSF PR 03-107) September 29, 2003.

[18] NSF Fact Sheet. From Supercomputing to Teragrid, September 2003.

Appendix D

Approval Processes in Other Agencies and Other Countries

This appendix begins with a set of tables that provide a comparative summary of the process used to set priorities for large research facilities for the most relevant unit of comparison at NSF, DOE, and NASA: NSF's MREFC Account, DOE's Office of Science, and NASA's Office of Space Science. These tables summarize the command and advisory structure, the process used to identify projects, the project evaluation criteria, the process used to involve the scientific community, the prioritization process, and the use of strategic planning in each process.

The tables are followed by several in-depth descriptions of features of the planning and approval process in use at different institutions. Included are a detailed presentation of the process at NSF, the strategic planning process at NASA's Office of Space Science, the approval and funding process at DOE's Office of Science, the 20-year facilities outlook activity at DOE's Office of Science, a discussion of the selection process for NSF Science and Technology Centers and NSF Engineering Research Centers, a description of DOD's Office of the Director for Defense Research and Engineering, and strategic planning and prioritization processes in use by the United Kingdom and Germany.

NSF

Relevant unit for agency comparison	**MREFC account**; annual budget about $150 million
Command structure	The director presides over seven program directorates spanning different fields of science and engineering. Additional offices provide administrative, financial, and management support. The director is advised by several NSF internal advisory councils of staff. As chartered, the National Science Board is the agency's governing board that establishes policies, oversees strategic planning, approves new programs and major awards, and oversees the general operations of NSF.
Advisory structure	Each directorate is advised by an advisory committee of external experts. Committees of visitors are used on a periodic basis to review and improve program operations within each directorate.
Origin of projects	Nominally, all projects come from the community. Large facility project ideas can be identified at community meetings, NSF-sponsored workshops, or by NSF program managers.
Strategic planning	NSF does not operate as a mission agency; rather, it strives to act as a facilitator for innovation and creativity. As required by GPRA the agency produces a 5-year strategic plan, last written in 2003. NSF does not regularly engaged in roadmapping activities across directorates. New standards are now requiring MREFC projects to provide life-cycle cost and management schedules. However, NSF's strategic plan does identify cross-cutting themes to attract attention and encourage fields of research. For instance, the "People, Ideas, Tools" theme of NSF plays a large role in directing its programs. Also, NSF identifies "cross-cutting investment areas," which are selected for substantial investment over the next few years.
Project evaluation criteria	• Need for such a facility. • Research that will be enabled. • Readiness of plans for construction and operation. • Construction budget estimates. • Operations budget estimates.
Community involvement	The NSF internal champion for a project is typically program staff for some sector of the scientific community. At the simplest level, MREFC projects arise from solicited proposals. Once a project has started to gain momentum, the channels for community input in project development are not standard, but typically involve workshops.
Prioritization process	Projects are recommended to the director by staff for consideration by the NSF internal MREFC review panel. The MREFC review panel evaluates the merit of a proposed project and then prioritizes it relative to other projects under consideration. The review panel and the director place particular emphases on the following criteria to determine the priority order of the projects:

- How "transformative" is the project? Will it change the way research is conducted or alter fundamental science and engineering concepts or research frontiers?
- How great are the benefits of the project? How many researchers, educators and students will it enable? Does it broadly serve many disciplines?
- How pressing is the need? Is there a window of opportunity? Are there interagency and international commitments that must be met?

The director then selects projects for review by the NSB Committee on Programs and Plans. The board authorizes the director to seek funding for a project in a future funding request. At the discretion of the director, projects are included in the agency's budget request.

DOE

Relevant unit for agency comparison	**Office of Science** (SC); annual budget about $3.3 billion, of which about 25 percent ($825 million) goes to major user facilities
Command structure	The SC director presides over six offices separated by scope of scientific research and one workforce development program. Additional offices provide management and administrative support. The director makes the final decisions regarding projects and planning on the basis of detailed documentation provided by each of the six program offices and their advisory committees. Because SC operates many large facilities, a recent reorganization has provided a specific Office of Engineering and Construction Management that centralizes project planning and oversight. In contrast with NSF, the SC director does not have the final word; SC's activities take place in the larger context of the entire DOE.
Advisory structure	There are six program advisory committees, each FACA-chartered. Members are scientific experts in the fields appointed by the secretary of energy; the committees are staffed by SC personnel. The SC director periodically charges each advisory panel to perform an assessment or to generate recommendations. The FACA committees typically spin off subpanels to address specific tasks assigned by the director.
Origin of projects	Large facility projects are usually based at national laboratories and therefore span many different fields of science research. Members of the scientific community develop project plans, usually in conjunction with one of the national laboratories. The SC advisory committees often solicit community input and convene planning workshops to identify community needs.
Strategic planning	A strategic plan was produced according to GPRA for 2001. SC is developing an updated mission statement and strategic plan that promise to improve on the spirit of GPRA. The SC director regularly charges each advisory committee to generate a long-range plan containing priorities and recommendations. These then become templates for the direction of each program office. SC does not traditionally generate a roadmap, although the current director has recently completed an effort to produce a 20-year facilities outlook.

Project evaluation criteria	• Scientific or technical merit of the educational benefits of the project. • Appropriateness of the proposed method or approach. • Competence of personnel and adequacy of proposed resources. • Reasonableness and appropriateness of the proposed budget. • Other appropriate factors, established and set forth by SC in a notice of availability or in the specific solicitation.
Community involvement	Community involvement is substantial and focused at the advisory committee level. They fall under FACA, so all full advisory-panel meetings are open to the public, encouraging an open process and public comment. Commissioned long-range plans involve substantial community input, often with town meetings, series of workshops, and calls for project ideas.
Prioritization process	SC relies on the expert opinions of the community to set priorities, particularly through the use of the independent advisory committees and extensive peer-reviewed competition for grants and contacts. Inputs to priority-setting process by the advisory panels include scientific opportunity, projected investment opportunity, DOE mission needs, and administration and departmental priorities.

NASA

Relevant unit for agency comparison	**Office of Space Science (OSS)**; annual budget about $4 billion
Command structure	The OSS Associate Administrator presides over four, mostly exclusive divisions focusing on different types of space science. The Associate Administrator makes final planning decisions on the basis of community and advisory committee inputs.
Advisory structure	Associate administrator is advised by the Space Science Advisory Committee (SScAC), a FACA-chartered committee. Smaller non-FACA committees representing each division (theme) report to SScAC. These committees are composed of science and engineering experts in the field and generally provide tactical advice. Strategic advice can be provided by independent bodies such as the NRC's Committee on Astronomy and Astrophysics or specific blue ribbon panels.
Origin of projects	Large missions are ranked by external working groups (often the National Research Council) over long timeframes; OSS issues research solicitations for specific missions and instrumentation to further develop plans and opportunities.
Strategic planning	A 5-year strategic plan for the entire agency is updated every 3 years (GPRA) with input and synthesis at each level. A combination of top-down vision and bottom-up mission proposals produces an agency-wide plan that is coherent and responsive. OSS develops a formal Strategic Plan that includes a roadmap and a detailed rationale for the set of projects recommended over the next 5 years and beyond.

Project evaluation criteria	• Scientific or technical merit, including competence. • Relevance to NASA's objectives (defined in planning process). • Realism and reasonableness of cost and management plans.
Community involvement	Origin of proposals, priority-setting procedures at division level, and advisory committees at all levels. The research community is an integral component of the strategic planning process that produces a set of science priorities that are then incorporated into the eventual roadmap.
Prioritization process	Division level: often outsourced to the community. Office level: community participation in "shootout" among division priorities under auspices of SScAC with roadmapping and budget input from NASA personnel.

NATIONAL SCIENCE FOUNDATION PRIORITY-SETTING PROCESS FOR MREFC ACCOUNT PROJECTS

Introduction[1]

About one-fifth of the NSF $5.3 billion budget in FY 2003 supports the development and provision of "tools," which are intended to provide what NSF calls "a widely accessible, state-of-the-art science and engineering infrastructure." Large facility projects are funded through the Major Research Equipment and Facilities Construction (MREFC) account and through other accounts encompassed in the tools budget category. The MREFC account represents about one-eighth of the Foundation's proposed investment in tools in FY 2003, rising to about one-sixth in the budget estimates for FY 2004. Despite representing a relatively small portion of the total NSF budget, the large facility projects supported through the MREFC account are highly visible because of their size and geographic concentration, and many of the issues raised by these projects must also be considered in other NSF projects and programs.

The large facility projects supported by NSF are nearly as varied as the scientific research that the Foundation supports. Some facilities represent new and increasingly powerful versions of instruments that have been used for decades to study the natural world, such as telescopes or particle accelerators. Other large facilities use new ways of gathering information; examples are a new facility designed to measure gravity waves generated by such cosmic events as star collisions and supernovae

---

[1]The text for this section has been reviewed and modified by NSF to reflect its practices as accurately as possible.

and a proposed facility that would detect high-energy neutrinos in a large volume of Antarctic ice to provide information about the astrophysical sources of extremely high energy cosmic rays. Some large facilities primarily serve specific scientific disciplines, such as optical telescopes and radio-telescopes for astronomy and observatory networks for oceanography. Other facilities enable research in a wide array of disciplines. For example, the ground facilities, ships, and aircraft stationed in Antarctica allow scientists to study the atmosphere, ice, oceans, and geology of the region.

Regardless of their detailed characteristics, all large facility projects are being affected by the accelerating development of information technologies. Increasing quantities and varieties of information are being gathered, rapidly analyzed, and interpreted. Information technologies are also changing the fundamental nature of many large facility projects. New information technologies are making it possible, for example, for many large facilities to consist of smaller instruments and research projects in widely distributed geographic locations. The George E. Brown, Jr. Network for Earthquake Engineering Simulation, which is intended to improve the seismic design and performance of the U.S. civil and mechanical infrastructure, will consist of 15 experimental equipment sites linked by a high-performance Internet system. Elements of EarthScope, a distributed project to study the structure and dynamics of North America, will operate in nearly every county in the United States during the project's eight- to ten-year lifetime. The proposed National Ecological Observatory Network will consist of geographically distributed observatories linked to laboratories, data archives, and computer modeling facilities.

Origins of Concept and Development of Proposals[2]

The origins of large facility projects are as varied as the projects themselves. Some arise as logical outgrowths of previous research or facilities. Others originate as a consequence of new scientific development where the need for a new facility becomes apparent where no such need existed before. In some cases, such as the provision of high-speed networks and computers, a large facility is required to enable other kinds of research. Other large facilities are focused on the acquisition of data that cannot be obtained in any other way.

The impetus for all new large facility projects originates within the scientific community, but ideas take various routes to fruition. The community processes vary greatly from field to field. Often, self-organizing groups within a field of science or engineering develop the initial ideas

---

[2]The text for this section has been reviewed and modified by NSF to reflect its practices as accurately as possible.

for a new facility and set scientific objectives for the facility by prioritizing competing needs. At other times, facilities have been proposed at the initiative of an individual scientist or a small group of researchers with a bold vision. NSF Program Officers and staff foster these initiatives by providing funds for meetings and workshops that facilitate the scientific community's internal evaluation and maturation of these concepts. In every case, the mission of the NSF is to seek out the best ideas and the best scientists and to empower their investigations.

This process of nurturing and maturation of a concept for a facility can take many years to fully develop, or it can come together as a funded proposal quite quickly, depending on the nature of the proposal, the immediacy of scientific need, and the potential payoffs scientifically and for society in general. The NSF's role in this process is reactive and responsive to the scientific community, rather than prescriptive, insuring that the highest quality proposals, as determined by peer review within the scientific community, are brought forward for implementation. NSF Program Officers are the key people who make the requirements for approval of such projects clear to the community.

In identifying new facility construction projects, the science and engineering community, in consultation with NSF, develops ideas, considers alternatives, explores partnerships, and develops cost and timeline estimates. By the time a proposal is submitted to NSF, these issues have been thoroughly examined.

Establishing Priorities for Large Projects[3]

Upon receipt by NSF, large facility proposals are first subjected to rigorous external peer review, focusing on the criteria of intellectual merit and the broad (probable) impacts of the project. Only the highest rated proposals, i.e., those that are rated outstanding on both criteria, survive this process. These are recommended for further review

- by an MREFC Panel that comprises the assistant directors and office heads, serving as stewards for their fields and chosen for their breadth of understanding, and chaired by the deputy director acting in consultation with the director; and subsequently
- by the National Science Board.

Both the MREFC Panel and the National Science Board look for a consistent set of attributes in projects they recommend:

---

[3]The text for this section has been reviewed and modified by NSF to reflect its practices as accurately as possible.

- The project represents an exceptional opportunity to enable frontier research and education.
- The impact on a particular field of research is expected to be transformational.
- The relevant research community places a high priority on the project.
- The resulting facility will be accessible to an appropriately broad user community.
- Partnership possibilities for development and operation are fully exploited.
- The project is technically feasible and potential risks are thoroughly addressed.
- There is a high state of readiness to proceed with development, in terms of engineering cost-effectiveness, interagency and international partnerships, and management.

The MREFC review panel evaluates the merit of a proposed project and then prioritizes it relative to other projects under consideration. It first selects the new projects it will recommend to the director for future NSF support, based on a discussion of the merits of the science within the context of all sciences that NSF supports. Using these criteria, projects that are not highly rated are returned to the initiating directorates and may be reconsidered at a future time. Then, highly rated projects are placed in priority order by the panel in consultation with the NSF director. The review panel and the director place particular emphases on the following criteria to determine the priority order of the projects:

- How "transformative" is the project? Will it change the way research is conducted or alter fundamental science and engineering concepts or research frontiers?
- How great are the benefits of the project? How many researchers, educators and students will it enable? Does it broadly serve many disciplines?
- How pressing is the need? Is there a window of opportunity? Are there interagency and international commitments that must be met?

These criteria are not assigned relative weights because each project has its own unique attributes and circumstances. For example, timeliness may be crucial for one project and relatively unimportant for another. Additionally, the director must weigh the impact of a proposed facility on the balance between scientific fields, the importance of the project with respect to national priorities, and possible societal benefits.

After considering the strength and substance of the MREFC Panel's recommendations, the balance among various fields and disciplines, and other factors, the director selects the candidate projects to bring before the NSB for consideration. The NSB reviews individual projects on their merits and authorizes the Foundation to pursue the inclusion of selected projects in future budget requests. In August of each year, the director presents the priorities, including a discussion of the rationale for the priority order, to the NSB, as part of the budget process. The NSB reviews the list and either approves or argues the order of priority. As part of its budget submission, NSF presents this rank-ordered list of projects to OMB. Finally, NSF submits a prioritized list of projects to Congress as part of its budget submission.

Project Implementation and Oversight[4]

Except for its facilities in the Antarctic, NSF does not directly operate research facilities. Rather, it makes awards to other organizations, such as universities, consortia of universities, or nonprofit organizations, to construct, operate, and manage the facilities. NSF enters into partnerships with those organizations, the details of which are most often defined through cooperative agreements, to accomplish this. The cooperative agreement defines the scope of work to be undertaken by the awardee and establishes the project-specific terms and conditions by which the NSF will maintain oversight of the Project. NSF has the final responsibility for oversight of the development, management, and performance of the facilities.

Each large facility project supported by NSF has a program manager in NSF who is the primary person responsible for all aspects of project oversight and management of the project within the foundation. The program manager carries out these responsibilities in accordance with an internal management plan (IMP) that has been crafted specifically for this project. The IMP defines a project advisory team (PAT) that consists of NSF personnel with expertise in the scientific, technical, management, and administrative issues associated with the project. The team works with the program manager to ensure the establishment of realistic cost, schedule, and performance goals for the project. The team also helps to develop terms and conditions of awards for constructing, acquiring, and operating a large facility. The NSF's director for large facility projects works closely with the program manager, providing expert assistance on non-scientific and non-technical aspects of project planning, budgeting, implementation, and management to further strengthen the oversight capabilities of the foundation. The deputy also facilitates the use of best

---

[4]The text for this section has been reviewed and modified by NSF to reflect its practices as accurately as possible.

management practices by fostering coordination and collaboration throughout NSF to share application of lessons learned from prior projects.

The awardee designates one person to be the project director. This person has the overall control and responsibility for the project within the awardee organization. Throughout the Implementation stage, the awardee executes and manages the project—either construction or acquisition—in accordance with the cooperative agreement between the awardee institution and the NSF. This phase of the project includes all installation, testing, commissioning, and acceptance. Oversight by the NSF during this phase is accomplished through periodic reviews, written reports by the awardee to the foundation that include documentation of technical and financial status using "Earned Value" reporting methods, annual work plans, periodic external reviews, and site visits.

By the end of the implementation stage, a proposal is submitted for operations and maintenance to the program manager. The program manager reviews proposals in accordance with the merit review procedures contained in Chapter V of the NSF's *Proposal and Award Manual* and presents a recommendation for funding to his or her division director and assistant director—office head. The Director's Review Board (or DRB) reviews proposals for awards exceeding the Director's Review Board threshold (see Chapter VI of the NSF's *Proposal and Award Manual*). Following DRB, the NSF director recommends awards above the National Science Board (NSB) threshold for approval to the NSB. The NSB reviews and approves awards recommended by the director. Following this, the assistant director—office head, through the division director, authorizes the program manager to recommend the making of an award in accordance with the proposal processing procedures contained in Chapter VI of the *Proposal and Award Manual.*

The program manager, with the Division of Grants and Agreements, drafts the cooperative agreement that will govern the project in accordance with the procedures contained in Chapter VIII of the *Proposal and Award Manual.* The Division of Grants and Agreements makes the award once the cooperative agreement is executed by it and the awardee.

NSF Director and National Science Board[5]

Using the recommendations received from the MREFC Panel, the NSF director selects candidate projects to be considered by the NSB during one of its meetings in the year. According to the *Guidelines*, the director uses the following criteria in making this selection:

---

[5]The text for this section has been reviewed and approved by the National Science Board.

- Strength and substance of the information provided to the MREFC Panel.
- The relationship to NSF goals and priorities, including NSF's educational mission.
- Appropriate balance among various fields, disciplines, and directorates, based upon a consideration of needs and opportunities.
- Guidance from the NSB on overall decision boundaries for the MREFC account, provided at the annual MREFC planning discussion (May).
- Opportunities to leverage NSF funds.

The NSB's Committee on Program and Plans (CPP) takes the lead in reviewing the proposed project; a member of the committee leads the discussion. The criteria considered by CPP are these:

- Need for such a facility.
- Research that will be enabled.
- Readiness of plans for construction and operation.
- Construction budget estimates.
- Operations budget estimates.

After the CPP reviews the project, it makes recommendations to NSB for approving its inclusion in future budget requests and for approving actual project implementation.

NSF Director and Office of Management and Budget

Once the NSB has approved a project for funding, the director may recommend the project for inclusion in a future budget request to OMB. In August of each year, the NSB reviews the NSF budget, which includes the list of projects being submitted to OMB for funding. For projects included in the budget request, a capital asset plan and justification must be prepared, following a format developed by OMB. The capital asset plan and justification provides a summary of how much the project will cost to build and operate, information on its management and cost, schedule, and performance goals and milestones.

The list of major projects in the budget may be modified during negotiations between OMB and NSF. During that process, other parts of the executive branch, such as the White House Office of Science and Technology Policy, may provide input on the projects included in the budget.

Congressional Action

After submission of the president's budget to Congress in February of each year, congressional subcommittees and committees examine the proposed expenditures and begin the appropriations process. Congressional appropriators make decisions about whether to fund each of the large facility projects proposed for NSF in the President's budget. In addition, because of budgetary constraints, the NSF director and OMB may decide not to request funds for large facility projects that the NSB has approved for inclusion in the budget.

By 2001, the NSB had approved six large facility projects that had not yet been funded. Concerns were expressed in Congress and elsewhere that political pressures rather than scientific merit would increasingly determine which projects received appropriations. In 2001, Congress asked NSF to rank the six projects in order of priority. NSF responded by dividing the projects into two categories of three projects each, with no ordering within a category. In its appropriations for FY 2003, Congress provided funds for two of the three projects in the high-priority category. In the 2004 budget request, NSF further ranked the projects, requesting funding for the remaining high-priority project in that fiscal year and proposing to start funding for the other three in FY 2005 and FY 2006.

The recent focus on NSF's setting of priorities among large facility projects continues a long-running discussion of the best way for NSF to support such undertakings. In the June 12, 2002, letter to NAS President Bruce Alberts that led to the present study, six senators stated that "funding requests by the Foundation for large facility projects appear to be ad hoc and subjective." The letter directed the National Academies to "review the current prioritization process and report to us on how it can be improved."

In the FY 2002 House conference report, Congress provided guidance as to the use of MREFC and R&RA expenditures. It stated that the purpose of the MREFC account is to provide resources for the acquisition, construction, and commissioning of large-scale research equipment and facilities, whereas the R&RA account is to fund planning, design, operations, and maintenance costs. Unless an exemption is granted, MREFC funding can no longer be used to fund planning and design costs, as has occurred in the past.

SELECTION PROCESS FOR NSF SCIENCE AND TECHNOLOGY CENTERS AND NSF ENGINEERING RESEARCH CENTERS

The NSF funds two programs for creating university-based research centers: The Science and Technology Centers (STC) and the Engineering Research Centers (ERC) programs. While the specific program goals of these two programs differ substantively, the overall review, renewal, and oversight schemes are similar. Both programs provide initial funding for five years with the possibility of extension to ten based on reviews performed during the initial funding period. Center selection follows an extensive review process wherein a program solicitation is sent to request preliminary proposals for consideration. After initial review, invitations are sent requesting full proposals from those groups whose pre-proposals meet review criteria. Additional review criteria are then imposed to decide the award recipients. The following describes specific details of the individual programs.

Science and Technology Centers

The NSF created the STC program in 1987 "to fund important basic research and education activities and to encourage technology transfer and innovative approaches to interdisciplinary programs" [1]. Since its inception, the program has funded four classes of centers in 1989, 1991, 2000, 2002, and the NSF released a program solicitation for preliminary proposals to the Class of 2005 in June 2003 [2, 3]. Twenty-three centers have completed the full 10 years of funding, and 11 centers currently operate on NSF funds. Two centers were closed prematurely due to management issues [2]. For consideration in the STC program, proposals must demonstrate cross-disciplinary research goals, an extensive education program, and a means of enabling knowledge transfer to industry or other interested parties such as government [2]. An example of an STC is the Science and Technology Center for Adaptive Optics at the University of California, Santa Cruz, which began funding in FY2000. The center brings together astronomers and vision scientists "to develop new instruments optimized for adaptive optics" with applications as diverse as imaging planets around nearby stars and 3-D construction of optic nerve fibers [4].

STC awards are made following a multistage review process. A program solicitation announces the request for preliminary proposals. These proposals are reviewed by "panels of individuals intellectually distinguished in their fields and experienced in integrative science, mathematics, engineering and technology research" [3]. The panels examine the preliminary proposals based on the merit review criteria common to all NSF programs:

- Intellectual merit.
- Broader impacts.
- Integration of research and education.
- Integration of diversity.

The preliminary proposals must also address three STC-specific review criteria:

- The value-added from funding the activity as a center.
- The efficacy of the proposed leadership and management plans.
- The integrative nature of the proposed center.

Groups whose preproposals demonstrate the most promise are then invited to submit full proposals. In the award selection process for the Class of 2005, 159 preproposals resulted in 37 invitations for full proposals [2]. The full proposals are again judged by mail and panel review according to the same criteria but with a special emphasis on the integrative nature of the center [3]. A subselection deemed most "worthy" then undergo site visit review where added emphasis is placed on the proposed management and leadership plan. An external ad hoc STC Advisory Committee makes a priority list of recommended centers based on the above criteria, "the potential national impact and legacy of the proposed activity, the balance of awards among scientific fields, geographical distribution, and the combined ability of the proposed Centers to meet the objectives of the STC Program." NSF management uses the list to make funding recommendations to the NSF director and the Director's Review Board. According to Bruce Umminger, senior scientist in the NSF Office of Integrative Activities (OIA), the stringent review process reflects the award size, $1.5 million–4 million/year, and the high visibility of the centers [2].

Following the award, centers continue to receive NSF oversight through their respective NSF directorates in coordination with the OIA. During its first 5 years of funding, each center undergoes annual reviews on which support for the following year is contingent. An in-depth review in the fourth year determines whether the center can renew for an additional 5 years of funding. The renewal request is evaluated through an ad hoc mail review and a formal on-site visit. In the event that a center is not renewed, additional funding at a decreased level is provided for 1 year. After a successful fourth-year review and renewal, centers continue to undergo NSF review at least every 18 months through the reduced, phase-out period of funding in years 9 and 10.

Engineering Research Centers

NSF created the ERC program in 1985 "to develop a government-industry-university partnership to strengthen the competitive position of U.S. firms in world trade and change the culture of engineering research and education in the U.S." [5]. The most recent crop of centers were announced October 3, 2003, bringing the number of engineering research centers currently under NSF support to 24 [6]. Before the announcement of the four new centers in 2003, the program had supported the creation of a total of 37 new centers over its lifetime [7]. Three elements structure the backbone of the ERC program: cross-disciplinary and systems-oriented research, education and outreach, and industrial collaboration and technology transfer [8]. The technological emphases of the current centers include bioengineering; design, manufacturing, and product development systems; earthquake engineering; and microelectronic systems and information technology. Among the newly named ERCs is the Engineering Research Center for Extreme Ultraviolet (EUV) Science and Technology (EUV ERC), headquartered at Colorado State University. EUV ERC will "explore the interface of physics, electrical engineering, chemistry, and biology using [EUV] light." The center already has begun partnerships with members of the semiconductor, laser, and advanced optics industries [6].

Similar to the STC program, ERC awards result from an extensive review process. Preliminary proposals are solicited through a program announcement inviting prospective teams to "develop a ten-year vision for advances in an emerging, potentially revolutionary or transforming engineered system" [7]. These preproposals are then reviewed by a panel of outside experts. At the time of submission, proposers are invited to suggest names of appropriate or inappropriate reviewers. When selecting reviewers, NSF places extra emphasis on finding persons from outside academe, from minority-serving institutions, and from related disciplines. The preproposals are evaluated according to the standard NSF criteria (listed above in the STC description) and a set of ERC-specific criteria [7]:

- Potential of the proposed engineered system to spawn new industries; transform the industrial base, service delivery system, or infrastructure; and have societal impact.
- Research plan that targets critical systems goals and challenging scientific and technical barriers and proposes projects to address them.
- Indication of an extensive understanding of the state of knowledge and the state of the art.
- An education plan that includes curriculum development at all levels.
- An outreach program that reaches a broad spectrum of faculty, teachers, and students.
- Convincing rationale for industrial partners and plan for including them in all aspects of the project.
- Appropriate institutional configuration that is well integrated across institutions in the case of multiuniversity centers.
- Available expertise to address all aspects of center research and

capable leadership, faculty, and students representing a diverse mix of sex, race, and ethnicity.

- An effective organizational structure and management plan.
- Necessary equipment, facilities, and laboratory space.
- A commitment to the interdisciplinary, educational, and diversity-building goals of the ERC program.

After initial review, full proposals are invited from a smaller selection of proposers. The full proposals are then reviewed according to additional criteria that include [7]:

- Proposed center space that can encourage interdisciplinary collaboration and house center management.
- Commitments from industry to become fee-paying members of the center.
- Industrial agreements that indicate a centerwide collaboration rather than a collection of individual projects and that facilitate technology transfer between the center and industry [7].

Reviewers are asked to submit a summary rating and recommendation on whether to fund each proposal. An NSF program officer assigned to each proposal uses the reviewers' advice to formulate a recommendation. These recommendations are sent to the Division of Engineering Education and Centers (EEC), where the director decides whether to accept the recommendation. The recommended proposals are forwarded to the Division of Grants and Agreements, where they are reviewed for business, financial, and policy implications, and the final funding decision is made [7].

After the initial award, all ERCs continue to receive NSF oversight and review. Each ERC is required to submit an annual report of progress and plans, and members of each ERC's leadership attend an annual meeting in Washington, D.C., to discuss progress, receive updates, and provide advice on the program. ERCs must also collect and submit progress indicator data to NSF. Like the STC program, the first ERC award provides funding for 5 years, with awards of up to $2.5 million/year. The annual reports submitted by the centers undergo outside merit review that forms the basis for determining funding levels for the following year. During either the third or sixth years, a center may submit a proposal for renewal to extend the award to 10 years. In the event that such a proposal does not receive approval, funding is phased out over a 2-year period "to protect the graduate students" [7]. A center that successfully applies for renewal will received decreased funding in years 9 and 10 to facilitate the transition to self-sufficiency [7].

STCs Initiated in FY2002

Center for Advanced Materials for Water Purification, Urbana, IL
   University of Illinois at Urbana Champaign (sponsor), Stanford University, and Clark Atlanta University

Center for Biophotonics Science and Technology, Sacramento, CA
   University of California, Davis (sponsor), University of California, San Francisco, University of California, Berkeley, Stanford University, and the Lawrence Livermore National Laboratory

Center for Embedded Network Sensing, Los Angeles, CA
   University of California, Los Angeles (sponsor), University of Southern California, University of California, Riverside, the NASA Jet Propulsion Laboratory, California State University, Los Angeles, and the California Institute of Technology

Center for Integrated Space Weather Modeling, Boston, MA
   Boston University (sponsor), Alabama A&M University, Dartmouth College, Rice University, Stanford University, University of California, Berkeley, University of Colorado, Boulder, University of Texas, El Paso, National Center for Atmospheric Research, NOAA Space Environment Center, and Science Applications International Corporation

Center on Material and Devices for Information Technology Research, Seattle, WA
   University of Washington (sponsor), University of Arizona, California Institute of Technology, University of Southern California, University of California, Berkeley, and University of California, Santa Barbara

National Center for Earth-surface Dynamics, Minneapolis, MN
   University of Minnesota, Twin Cities (sponsor), Fond Du Lac Tribal and Community College, Massachusetts Institute of Technology, University of California, Berkeley, Princeton University, and Science Museum of Minnesota

ERCs Awarded in 2003

Engineering Research Center for Extreme Ultraviolet Science and Technology (EUV ERC), Fort Collins, CO
   Colorado State University (headquarters), University of Colorado at Boulder, University of California, Berkeley, and Lawrence Berkeley National Laboratory

Engineering Research Center for Environmentally Beneficial Catalysis (CEBC), Lawrence, KS
   University of Kansas in Lawrence (headquarters), University of Iowa in Iowa City, and Washington University at St. Louis

Engineering Research Center for Collaborative Adaptive Sensing of the Atmosphere (CASA), Amherst, MA

University of Massachusetts at Amherst (headquarters), Colorado State University, University of Oklahoma, University of Puerto Rico at Mayaguez, National Severe Storms Laboratory, Oak Ridge National Laboratory, and Massachusetts Department of Education

Engineering Research Center for Biomimetic Microelectronic Systems (BMES), Los Angeles, CA

University of Southern California (headquarters), California Institute of Technology, and University of California, Santa Cruz

References

[1] NSF Fact Sheet on Science and Technology Centers: Integrative Partnerships.
[2] National Science Board meeting discussion by Bruce Umminger, October 16, 2003.
[3] NSF Program Solicitation for STC (NSF 03-550), March 3, 2003.
[4] NSF Press Release (NSF PR99-45), July 29, 1999.
[5] Report from the Engineering Education and Centers Division of the NSF Directorate for Engineering, *The Engineering Research Centers (ERC) Program: An Assessment of Benefits and Outcomes*, December 1997.
[6] NSF Press Release (NSF PR03-115), October 3, 2003.
[7] NSF Program Solicitation for ERC (NSF 02-24), November 29, 2001.
[8] NSF website on ERC, www.eng.nsf.gov/eec/erc/directory/erc_a.htm.

PROJECT APPROVAL PROCESS AT DEPARTMENT OF ENERGY OFFICE OF SCIENCE

The Department of Energy's Office of Science (SC) has a long history of initiating and supporting large-facility projects. Although the success of individual project management has been varied, the current DOE guidelines provide a robust framework for the formal development of project ideas.

Large Facility Project Procedures

On the basis of input from the SC advisory committees, staff perform an initial evaluation of all proposal applications to ensure that required information is provided, that the proposed effort is technically sound and feasible, and that the effort is consistent with program funding priorities. For applications that pass the initial evaluation, the office reviews and evaluates each application received on the basis of criteria set forth below and in accordance with the merit-review system. Evaluators are selected on the basis of their professional qualifications and expertise. They evaluate new and renewal applications according to the following criteria, listed in descending order of importance[6].

1. Scientific or technical merit or educational benefits of the project.
2. Appropriateness of the proposed method or approach.
3. Competence of applicant's personnel and adequacy of proposed resources.
4. Reasonableness and appropriateness of the proposed budget.
5. Other appropriate factors, established and set forth by the office in a notice of availability or in a specific solicitation.

DOE considers, as part of the evaluation, other available advice or information and such program-policy factors as ensuring an appropriate balance among program subjects. In addition to the evaluation criteria, the recipient's performance under an existing award during the evaluation of a renewal application are considered. Applications are chosen for award on the basis of the findings of the technical evaluations, the importance and relevance of the proposed application to the office's mission, and funds availability. Cost reasonableness and realism are also considered to the appropriate extent.

For projects over $5 million, the procedure is more formal. Special guidelines from DOE's Office of Engineering and Construction Management apply, and the Office of Science's Construction Management Support Division is more directly involved. Projects are tracked by their progress along so-called critical decisions (CDs). A CD is a formal determination or decision at a specific point in a project phase that allows the project to proceed to the next phase and commit resources. CDs are required during the planning and execution of a project, for example, before commencement of conceptual design, commencement of construction, or start of operations. CDs for traditional construction projects include the following[7]:

- CD-0, Approve Mission Need
  - Authorizes use of program funds for conceptual design studies.
  - Requires preconceptual planning document.
  - Requires mission-need justification document and external independent review.
- CD-1, Approve Preliminary Baseline Range
  - Allows expenditure of project engineering and design funds for design work.

---

[6]FY 2004 Congressional Budget Request for DOE, Science, Basic Energy Sciences, p. 256.
[7]DOE Order O413.3. October 2000.

- ○ Requires preliminary acquisition, project-execution, and risk-analysis plans.
- CD-2, Approve Performance Baseline
  - ○ Establishes baseline budget for construction.
  - ○ Continues design process.
  - ○ Authorizes development of request for construction funding.
  - ○ Requires review of contractor management system.
  - ○ Requires external independent review of performance baseline.
- CD-3, Approve Start of Construction
  - ○ Approves expenditure of funds for construction.
  - ○ Requires readiness of final design and procurement packages.
  - ○ Requires external independent review of execution readiness.
- CD-4, Approve Start of Operations or Project Closeout
  - ○ Allows start of operations or closeout of project.
  - ○ Includes readiness review and acceptance report.

Advisory Committees

To ensure that resources are allocated to the most scientifically promising experiments, DOE and its national laboratories seek external input by using a variety of advisory bodies. The FACA-chartered advisory committees provide advice to DOE on a continuing basis regarding the direction and management of the national energy research program. SC comprises six science offices and the associate director of each office is advised by its own corresponding advisory committee. The advisory committees meet regularly to advise the sponsoring agencies (for instance, the High Energy Physics Advisory Panel is jointly sponsored by DOE and NSF) on their research programs, assess their scientific productivity, and evaluate the scientific case for new facilities. Each advisory committee solicits input from the community during its regular long-range planning exercises. The call for project solicitations is made at meetings of professional societies and usually at community workshops that are considering related issues.

The Office of Science has six program advisory committees, each FACA-chartered. Each advisory committee provides valuable, independent advice to DOE on the complex scientific and technical issues that arise in the planning, management, and implementation of its program. Recommendations include advice on establishing research and facilities priorities, determining proper program balance among disciplines, and identifying opportunities for interlaboratory collaboration, program integration, and industrial participation. The committee includes mainly representatives of universities, national laboratories, and industries involved in energy-related scientific research. Particular attention is paid to obtaining a diverse membership with a balance of disciplines, interests, experience, points of view, and geography. Members are appointed by the secretary of energy; the committees are staffed by DOE personnel.

The director and the associate directors periodically charge each advisory panel to perform assessments or to address particular questions. The committees typically convene ad hoc subpanels to respond to the charge; an ad hoc report is filed with the parent advisory committee on completion. The committee can adopt or modify the subpanel report, with an accompanying written justification, before forwarding its final recommendation to the commissioning office.

One of the most important functions of the advisory committees is the development of long-range plans that express communitywide priorities for research. The most recent such plan was submitted in January 2003 at the request of the director as part of the 20-year facilities roadmap initiative and presented a roadmap for each field, laying out the science opportunities that each planning subpanel could envision as possibilities for the next 20 years. Large facility projects first appear in the community, work their way into the frequent but irregular advisory committee long-range plans, and eventually undergo development, typically at one of the national laboratories to leverage existing resources and expertise.

DEPARTMENT OF ENERGY OFFICE OF SCIENCE TWENTY-YEAR FACILITY OUTLOOK PRIORITIZATION PROCESS[8]

The DOE's Office of Science began to prioritize future major facilities in the fall of 2002. The Associate Directors of the Office of Science[9] were asked to list major facilities required for world scientific leadership in their respective programs out to 2023. Each Associate Director was given a funding "envelope" under which they were to include their estimated research budgets as well as the major facility planning, construction, and operating costs.[10]

---

[8]The text for this section was contributed by the Office of the Director of DOE's Office of Science, and also appears in DOE Office of Science, *Facilities for the Future of Science: A Twenty-Year Outlook*, Washington, D.C.: U.S. Department of Energy, 2003, pp. 9-10.

[9]The Office of Science has an Associate Director for each of its scientific programs: Advanced Scientific Computing Research, Biological and Environmental Research, Basic Energy Sciences, Fusion Energy Sciences, High Energy Physics, and Nuclear Physics.

[10]These envelopes were constructed from the "Biggert Bill" authorization levels for the Office of Science for FY 2004 through FY 2008 (since replaced by H.R. 6 and S. 14), and then a 4 percent increase in authorization level each subsequent year until 2023. The Office of Science understands that construction of the facilities listed within the envelopes will depend on many factors, including funding being available as needed and all technology hurdles surmounted as planned. Nevertheless, the envelopes, and the facilities listed within them, are consistent with a far-reaching vision of how and when the Office could contribute to DOE's missions and the nation.

Forty-six facilities were identified and phased to conform to perceived scientific opportunities over this 20 year period. Internal hearings were held, with each Associate Director describing the nature of the recommended facilities, together with the scientific rationale behind their choices. This process was completed in December 2002.

The program prioritizations by the Associate Directors were then submitted in mid-January 2003 to the respective program's Advisory Committee, with a request for an analysis of the relative scientific opportunities associated with each of the facilities proposed by their respective Associate Director, and with any additions they felt important that may have been omitted. In some cases, the Committees were requested specifically to work together to capture the interdisciplinary needs that might be missed if a Committee focused too narrowly on its own traditional discipline. The Office of Science Advisory Committees are chartered to bring to each program the full breadth of perspectives of the U.S. scientific community. Of the 118 people that sit on the Office of Science Advisory Committees, 64 percent are from universities, 15 percent from DOE laboratories, 10 percent from industry, 3 percent from other government agencies, and 8 percent from other types of institutions.

The Advisory Committees recommended 53 major facilities for construction, and assessed each according to two criteria: scientific importance and readiness for construction. Against the first criteria, the committees divided their facilities into three categories: highest scientific importance, secondary scientific importance, and hard-to-assess scientific importance. The Committees also categorized the facilities into "near-term," "mid-term," and "far-term" according to their readiness for construction.

The results were plotted in a matrix illustrated in Figure D-1. "Highest scientific importance" was divided into categories A, B, and C, depending upon readiness for construction. "Secondary scientific importance" was labeled as category D, and "hard-to assess scientific importance" as category E.

With this input from the Advisory Committees, the challenge remained to prioritize the facilities across scientific disciplines.[11] The

---

[11]While prioritizing scientific programs and/or facilities within disciplines can be difficult, it is done regularly throughout the Federal Government and by numerous scientific and technical advisory committees. Prioritizing openly across disciplines, however, is notoriously difficult and has been done rarely. Physicist William Brinkman recently testified before the House Committee on Science to the effect that while such prioritizations are possible, they are necessarily based on intuition and therefore subjective. David Goldston, staff director for the Committee, responded that the Committee understood this and it was the reason that the Committee wanted "someone else" to do the prioritization. For further description of this discussion and the difficulties of prioritizing across fields see *Science and Engineering Infrastructure for the 21st Century, the Role of the National Science Foundation*, Draft December 2002, and 'National Science Foundation: Secrecy on Big Projects Breeds Earmarks, Panel is Told," Jeffrey Mervis, *Science*, Vol. 300, May 30, 2003.

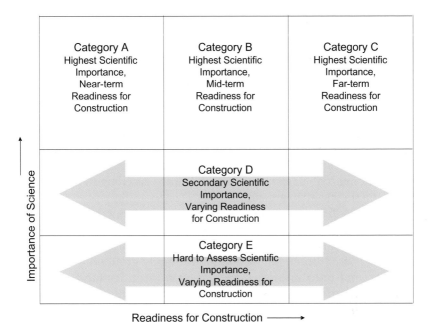

FIGURE D-1. DOE's Office of Science facilities matrix.

Director of the Office of Science addressed this challenge by prioritizing the 53 facilities according to his assessment of their scientific promise and their fit with the Department's missions. The costs associated with the Office of Science's base research programs and the other responsibilities were added, and the entirety was made to fit under an aggressive funding envelope (see footnote 2 in this appendix) extended through 2023. Twenty-eight projects survived, along with a contingency in the "out-years," recognizing the need for flexibility over a 20-year period.

A Benchmark for the Future

The *Twenty Year Outlook* represents a snapshot—the DOE Office of Science's best guess today at how the future of science and the need for scientific facilities will unfold over the next two decades. We know, however, that science changes. Discoveries, as yet unimagined, will alter the course of research and so the facilities needed in the future.

For this reason, the *Outlook* should be assessed periodically in light of the evolving state of science and technology. The *Outlook* will also serve as a benchmark, enabling an evaluation of facilities proposed in the future against those on this list. Future revisions should maintain the funding envelope used to guide this list, enforcing fiscal discipline upon discussions and requiring the elimination of facilities in order to accommodate more important or exciting prospects.

The DOE's Office of Science recognizes that the breadth and scope of the vision encompassed by these 28 facilities reflects a most aggressive and optimistic view of the future of the Office. Nevertheless, we believe that it is necessary to have and discuss such a vision. See Figure D-2. Despite the uncertainties, it is important for organizations to have a clear understanding of their goals and a path toward reaching those goals. The *Outlook* offers just such a vision.

The 28 facilities are listed by priority [above]. Some are noted individually; however others, for which the advice of the Advisory Committees was insufficient to discriminate among relative priority, are presented in "bands." In addition, the facilities are roughly grouped into near-term priorities, mid-term priorities, and far-term priorities (and color-coded red, blue, and green, respectively) according to the anticipated R&D timeframe of the scientific opportunities they would address.

Each facility listing is accompanied by a "peak of cost profile," which indicates the onset, years of peak construction expenditure, and completion of the facility. Because many of the facilities are still in early stages of conceptualization, the timing of their construction and completion is subject to the myriad considerations that come into play when moving forward with a new facility.

NATIONAL AERONAUTICS AND SPACE ADMINISTRATION OFFICE OF SPACE SCIENCE PLANNING PROCESS[12]

NASA's strength is in its strategic planning, which uses community input to formulate priorities in a public and independent process. As a mission agency, the Office of Space Science (OSS) sets goals and objectives that are developed through a careful process. Project proposals are measured against them and organized for development. On the front end, announcements of opportunity for particular types of projects are used to solicit proposals. These can be as general as a "dark energy probe" or as specific as a specific instrument set for a planned mission. On the implementation and operation end, NASA is relatively successful through the use of its subordinate but independent centers (typically

---

[12]The text for this section has been drawn from the FY 2000 Space Science Enterprise Strategic Plan, pp. 48-49.

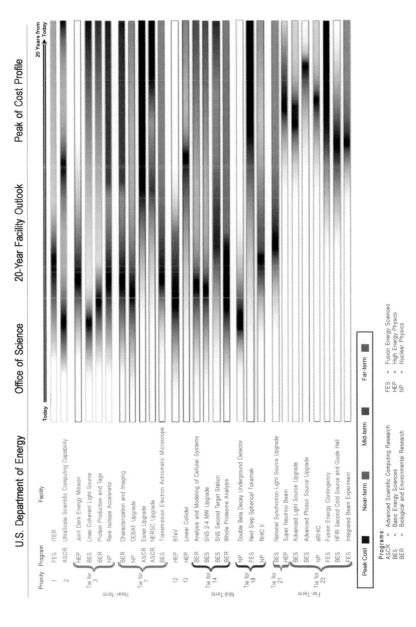

FIGURE D-2 DOE's 20-year outlook.

Goddard Space Flight Center for OSS). Finally, NASA projects are spear-headed by, essentially, principal investigators, who include a management plan in the proposal and work closely with the host center.

NASA engages in an agencywide strategic planning process every 3 years, producing an updated 5-year plan that includes a detailed roadmap accommodating the projected budget envelope. That process is designed to satisfy NASA's compliance with the Government Performance and Results Act of 1993 (GPRA), but it has become a powerful tool for identifying and shaping the mission of the agency, tying together budget, performance, and vision. To be synchronized with the triennial revision of the agency strategic plan mandated by GPRA, OSS revises its own strategic plan at the same interval. Each unit measures the consistency of its projects against the overall NASA objectives, the OSS objectives, and its own goals to produce a roadmap. The roadmaps are considered jointly by the enterprise at a closed but broadly representative retreat where determinations about relative importance and priority are made. OSS then drafts its strategic plan; this involves another series of open planning meetings that synthesize the different roadmaps, the broader goals of the enterprise, and additional information about budget expectations. NASA headquarters uses each enterprises's strategic plan to create the final and authoritative NASA strategic plan.[13] This final document becomes the basis for the agency's annual budget request for the next few years.

The NASA vision communicates the agency's mandate in the 21st century. The NASA mission lays out a clear path to the future. The mission provides a framework for developing goals that each unit of NASA must achieve. NASA has seven strategic goals (expanding on the mission) that enable each enterprise to focus planning, manage programs, and measure results. Each of the agency's six enterprises uses the strategic goals to define its programs. The agency goals are further broken down into enterprise objectives. (These are, essentially, projections of the agency's mission onto the subspace spanned by OSS.)

The strategic-plan development process depends on active involvement of outside parties, especially the space-science research community. The National Research Council (through its Board on Physics and Astronomy, Space Studies Board, and their discipline subcommittees) develops long-range strategic-program assessments and recommendations. The Space Science Advisory Committee (SScAC), on the basis of inputs

---

[13]The degree of top-down vs. bottom up planning at this stage varies, depending in part on NASA's leadership. For instance, the 2003 NASA strategic plan was released far in advance of the OSS strategic plan, forcing OSS to play the lead role in synthesizing the theme roadmaps with the overall NASA vision. However, the 5-year plan is revised every 3 years, so the overarching strategic plan is still based heavily on grass-roots priorities.

from its own subcommittees, provides the enterprise with roadmaps that integrate National Research Council and additional community inputs with technical, budget, and programmatic factors.

2000 Strategic Planning Process[14]

A narrative description of the process used to develop the FY 2000 Space Science Strategic Plan is an instructive example. This description is from the *The Space Science Enterprise Strategic Plan* published in November 2000; note that the Space Science Enterprise refers to the Office of Space Science.

> Work on the 2000 plan began in late 1998, when the Enterprise's Science Board of Directors initiated the development of science and technology roadmaps for each Enterprise science theme (Astronomical Search for Origins, Structure and Evolution of the Universe, Solar System Exploration, and Sun-Earth Connection). Those roadmaps—which were developed by roadmapping teams that included scientists, engineers, technologists, educators, and communicators of science—address science goals, strategies for achieving the goals, missions to implement the strategies, technologies to enable the missions, and opportunities for communicating with the public. Each roadmapping team was built from or overseen by its theme subcommittee of the Space Science Advisory Committee. The teams each held a series of meetings to obtain science priority views from community scientists, hear advocacy presentations for specific missions, examine technology readiness for alternative mission options, and discuss relative science priorities, balance, and optimal activity sequencing in light of this community input. One technique used to foster convergence was to take straw polls among team members during successive meetings.
>
> At the end of the roadmapping period, each of the four theme roadmapping activities submitted a summary document outlining science and mission recommendations to the Space Science Advisory Committee and to Enterprise Headquarters management. Enterprise management then combined the mission recommendations of the roadmapping teams into an integrated mission plan, guided by the current OMB 5-year budget profile, realistic estimates of most likely future resource availability beyond that, and additional agency-level and administration guidance. Likewise, science goals in the roadmaps were used to examine and restate those presented in the 1997 Enterprise Plan.
>
> An integrated roadmap was presented and discussed at a planning workshop that expanded the membership of the Space Science Advisory Committee with other community members and representatives of the

---

[14]*The Space Science Enterprise Strategic Plan*, NASA, November 2000, p. 48.

technology, education, and public-outreach communities. Attendees at the workshop also analyzed and revised the proposed updated science objectives and derived a new set of shorter-term research activities. The resulting consensus mission plan and goals, objectives, and research activities serve as the nucleus for the current Strategic Plan.

A draft of this Plan was provided to the Space Studies Board and its committees for review and feedback, and guidance received was used in finalizing the Plan. The findings and recommendations of the Academy's recently completed ten-year astronomy and astrophysics survey were consulted to assure consistency with the draft Plan. Finally, the Space Science Advisory Council had an opportunity to review the revised Plan and suggest changes before the Plan went to press.

The Space Science Enterprise strategic plan serves several purposes. It facilitates a consensus process in the science community that focuses on goals and priorities for the future. It serves OSS by providing a reference for programmatic decision making, and it provides input to the overall agency strategic plan and material to meet GPRA requirements. The strategic plan becomes a tool for use by OMB and Congress in program and budget advocacy. Finally, it provides a handbook for the public on what space science is going to do and why.

DEPARTMENT OF DEFENSE OFFICE OF THE DIRECTOR OF DEFENSE RESEARCH AND ENGINEERING

The Department of Defense (DOD) Science and Technology (S&T) Program supports the fundamental research, development, and demonstrations in science and technologies identified as important to military capabilities and operations. The extent of the S&T Program includes the development of the nation's high technology weapons systems, the technology base upon which those system rely, and the equipment to both support and prepare military personnel. The program also plays an important role in developing certain key technologies that transfer to commercial applications and help to grow the overall economy. Basic research (6.1), applied research (6.2), and advanced technology development (6.3) together comprise DOD's Science and Technology (S&T) program. S&T projects seek new ways of accomplishing tasks of military value and the underlying scientific and engineering principles involved.

Within DOD, the Office of the Director of Defense Research and Engineering is primarily responsible for the basic research plan of the department, "to ensure that the warfighters today and tomorrow have superior and affordable technology to support their missions, and to give them revolutionary war-winning capabilities."

Project Selection

Project proposals solicited by the Directorate of Defense Research and Engineering are typically judged on the following criteria:

- DOD mission priority.
- Military advantage gained by exploiting the requested S&T.
- Merit of scientific study.
- Potential for significant progress.
- Appropriateness of solution for meeting requirements.

Program Evaluation[15]

DOD uses defense technology objectives (DTO) to provide focus for the development of technologies that address identified military needs across the department. Each DTO identifies a specific technology advancement that will be developed or demonstrated, with expected date of availability, specific benefits resulting from it, and the amount of funding needed. The DTO process is used to comply with GPRA. The output of this process includes budget and management decisions.

The methodology used for evaluating the S&T program is known as technology area reviews and assessments (TARA). TARA is the department's official response to GPRA, and it is a mechanism to evaluate science and technology programs through expert peer reviews. The evaluation of basic and applied research is carried out by internal agency panels of experts and by TARA review panels. Each panel consists of 10-12 technical experts from academe, industry, and nonprofit research organizations. Most TARA team members are recognized experts from the National Academies, the Defense Science Board, the scientific advisory boards of the military departments, industry, and academe. Each is chaired by a senior executive appointed by the deputy under secretary for S&T.

These teams are asked to evaluate the programs for quality, advances in leading the state of the art in research areas, and for their scientific vision. The department requires that two-thirds of each panel be experts from outside DOD. One-third of each panel's members are refreshed at the time of each reviewing cycle.

Defense Science Board

The Defense Science Board operates by forming task forces consisting

---

[15]Portions of this section have been excerpted from *Implementing the Government Performance and Results Act for Research: A Status Report*, National Academy Press, 2000.

of Board members and other consultants and experts to address those tasks referred to it by formal direction. The products of each task force typically consist of a set of formal briefings to the board and appropriate DOD officials and a written report containing findings, recommendations, and a suggested implementation plan. The board reports directly to the secretary of defense through the under secretary of defense for acquisition, technology and logistics (USD [AT&L]) while working in close coordination with the DDR&E to develop and strengthen the department's research and development strategies for the 21st century. The board met for the first time on September 20, 1956. Its initial assignment examined basic research, component research, and technology advancement programs and their administration in S&T areas of interest to the DOD. On December 31, 1956, a charter was issued specifying the Board as advisory to the assistant secretary of defense (research and development). The subsequent consolidation of the offices of the assistant secretaries of defense for R&D and applications engineering in 1957 resulted in the board's reconstitution as adviser to the secretary of defense through the assistant secretary of defense (research and engineering).

The mission of the board is to advise, in response to formal requests, the secretary of defense, the deputy secretary of defense, the under secretary of defense for acquisition, technology and logistics, and the chairman of the Joint Chiefs of Staff on matters relating to science, technology, research, engineering, manufacturing, and acquisition process and other matters of special interest to the DOD. The board specifically concerns itself with pressing and complex technology problems facing DOD in such areas as research, engineering, and manufacturing. It seeks to ensure the identification of new technologies and new applications of technology in these areas for the strengthening of national security. The board does not advise on individual procurements.

Defense Advanced Research Projects Agency[16]

DARPA fulfills a unique role within the Department of Defense. As a Defense Agency, DARPA reports to the Secretary of Defense. The Director, Defense Research and Engineering has been assigned to be DARPA's Principal Staff Assistant (PSA). DARPA is the Secretary of Defense's only research agency not tied to a specific operational mission. DARPA supplies technological options for the entire Department. DARPA is designed

---

[16]Excerpted from *Strategic Plan Defense Advanced Research Projects Agency,* February 2003, p. 5. Available on the DARPA web site: (http://www.darpa.mil/body/pdf/ DARPAStrategicPlan2003.pdf).

to be the "technological engine" for transforming the Department of Defense.

DARPA's decision-making process is somewhat unusual for a government agency. It is informal, flexible, and yet highly effective because it focuses on making decisions on specific technical proposals based on the factors discussed above. There are two reasons for this. DARPA is a small, flat organization rich in military technological expertise. There is just one porous management layer (the Office Directors) between the program managers and the Director. With less than 20 senior technical managers, it is easy to make decisions. This management style is essential to keeping DARPA entrepreneurial, flexible, and bold. DARPA's management philosophy is to pursue fast, flexible, and informal cycles of "think, propose, discuss, decide, and revise." This approach may not be possible for most government agencies, but it has worked well for DARPA.

The Basic Process: DARPA uses a top-down process to define problems and a bottoms-up process to find ideas, involving the staff at all levels. DARPA's upper management and program managers identify "DARPAhard" problems by talking to many different people and groups. This process includes:

- Specific assignments
- Requests for help
- Discussions with senior military leaders
- Research into recent military operations
- Discussions with defense agencies
- Discussions with the intelligence community
- Discussions with other government agencies or outside organizations
- Visits to Service exercises or experiments.

During DARPA's program reviews, which occur throughout the year, DARPA's upper management looks for new ideas from program managers (or new program managers with ideas) for solving these problems. At the same time, management budgets for exploring highly speculative technology that have far-reaching military consequences.

Program managers get ideas from many different sources, such as:

- Their own technical communities;
- Suggestions from DOD-wide advisory groups, including the Defense Science Board and Service science boards;
- Suggestions from DARPA-sponsored technical groups
- Suggestions from industry or academia, often in response to published Broad Area Announcements or open industry meetings
- Breakthroughs in DARPA programs and/or U.S. or international

research

During reviews of both proposed and ongoing programs, DARPA's assessment is often guided by a series of questions. These seemingly simple queries help reveal if a program is right for DARPA.

- What is the project trying to do?
- How is it done now and what are the limitations?
- What is truly novel in the approach that will remove those limitations and improve performance? By how much?
  - If successful, what difference will it make??
  - What are the midterm exams required to prove the hypothesis?
  - What is the transition strategy?
  - How much will it cost?
  - Are the programmatic details clear?

SUMMARY OF UK RESEARCH COUNCILS' PRIORITY-SETTING PROCESS

On behalf of the research councils of the United Kingdom (RCUK), the UK Office of Science and Technology (OST) released the second edition of its *Large Facilities Strategic Roadmap* in June 2003.[17]   Several excerpts from that document are provided here. In 2001, the UK initiated a new process for treating proposals for large facilities. It begins with the production of a roadmap that contains projects currently identified as being of the highest strategic importance. There is no commitment to funding and no guarantee that the UK will participate in all the projects in the roadmap.

From Section 1, "Executive Summary":

It is a tool by which Research Councils UK (RCUK) and its members can assess strategically the most expensive and complex scientific facilities with which UK researchers are or may wish to be involved. The road-map includes facility "projects" identified by members of RCUK as a priority for consideration which meet one or more of the following criteria:

- Where there could be an international dimension to the proposed

---

[17]The excerpts provided here are taken from the UK Office of Science and Technology Web site, where the *Large Facilities Strategic Roadmap* is posted at http://www.ost.gov.uk/research/funding/lfroadmap/index.htm.

facility and therefore opportunity to share costs and develop relationships to benefit the UK science programme;

- Where the facility supports the requirements of research communities of more than one Research Council;
- Where the capital investment is greater than the sum of £25 million, when it represents a significant element of an individual Research Council's budget line.

The roadmap is scheduled to be updated every 2 years; the June 2003 version represents the second iteration; it contains projects in 10 strategic fields. The roadmap is the first stage of a gateway process that requires all large capital investments to be managed as discrete projects subject to review and independent scrutiny at all stages in their life cycle.

From Section 5, "OGC Gateway Process and How it is Used in Large Science Facilities":

The OST and members of RCUK use the Office of Government Commerce's Gateway process to help procure large scale scientific facilities. All new procurement projects in civil Central Government—including NDPBs—are subject to the Gateway process, which examines a project at critical stages in its lifecycle to provide assurance [that] it can progress successfully to the next stage. The process has a series of Gateway Reviews, as follows:

0. To confirm the overall strategic assessment
1. To confirm the business justification
2. To confirm the procurement method and sources of supply
3. To confirm the investment decision - before letting any contracts
4. To confirm "readiness for service"
5. To confirm "in service benefits"

In the case of scientific projects, the first stage must involve an independent assessment of scientific value, including some form of peer review that addresses the following criteria:

- Importance (depth) of science knowledge to be delivered by project.
- Breadth of science knowledge that will benefit from investment.
- Match with international positioning of UK science.
- Strength of opportunity for training (links to numbers of users).
- Contribution to or from UK technology-industry base.
- Opportunity for spin-off and exploitation.

See additional material in Appendix E concerning the criteria used in the strategic roadmapping process.

SUMMARY OF GERMAN MINISTRY FOR EDUCATION AND RESEARCH'S PRIORITY-SETTING PROCESS[18]

Germany has a distinctive science system, with responsibilities shared between federal and state governments. At the federal government level, the Ministry of Education and Research (BMBF) is the key player. Various nonprofit bodies play important roles in distributing funding; for example the Deutsche Forschungsgemeinschaft is a major source of project funding for universities, and the Max Planck Gesellschaft operates 81 research institutes.

National policy coordination and policy advice are provided by two joint federal-state bodies:

• The Bund-Länder-Kommission für Bildungsplanung und Forschungsförderung, which is a forum for discussions between federal and state ministries.

• The Wissenschaftsrat, or Science Council, an independent body whose members are appointed by the federal president and that advises federal and state governments on all matters of higher education and research policy. Its members include representatives of the science community and nominated representatives of state and federal governments.

Over the last 20 years, the BMBF has set up ad hoc groups to recommend priorities for funding large facilities that have been proposed by the scientific community whenever the need was felt (their reports have become known by the names of the chairs: Mayer-Leibnitz, Pinkau, and Grossman). Such recommendations go to the minister, who makes final decisions (subject to discussion with the Finance Ministry and at cabinet level when appropriate).

In 2000, the BMBF asked the Wissenschaftsrat (henceforth referred to as the Science Council) to review nine proposals for large facilities for basic research (each priced at at least 15 million euros); and in January 2001, the Science Council established a working group for this purpose. The group included scientists at universities and research establishments in Germany, the United States, and Switzerland and "individuals involved in and representing national and international scientific administrations." The group established six subpanels composed of 57 external experts, including 37 from abroad.

---

[18]This text has been drawn from the final report of the Science Council, *Statement on Nine Large-scale Facilities for Basic Scientific Research and on the Development of Investment Planning for Large-scale Facilities*, 2003. The report is also available on the Science Council's Web site at http://www.wissenschaftsrat.de/texte/5385-02.pdf.

The Science Council's working group not only evaluated specific projects but laid out a general framework for such work and proposed criteria. It seems likely that, as proposed in the report, the Science Council will be asked to carry out such reviews at regular intervals in the future, replacing the previous ad hoc procedure.

The council established a working group with six expert subpanels that assessed projects in particular fields according to the following criteria:

- The probability of fundamental new insights or the possibilities of decisive scientific advances that could be achieved only with the large-scale facility.
- The large-scale facility's technical feasibility and degree of technical innovation.
- The scientific and technical competence of the institutions involved.
- The already existing or anticipated acceptance of the (potential) users with immediately relevant and related expertise.
- The fulfillment of various objectives of importance for research (transfer, international perspectives, and promoting young scientists).

The working group then considered all the projects, taking into account:

- Scientific potential of the research program.
- Fulfillment of science and technology policy goals as formulated in 10 general "theses on the significance of large-scale facilities for basic scientific research."
- Degree of maturity of the technical concept and, connected to it, the possible timeframe for implementing the individual components.
- The context of further national and international scientific development of the research fields to which they belong and their interaction with other disciplines.

Finally, the working group divided the nine projects into three categories, the first meriting unconditional support, the second with specific points yet to be clarified, and the third requiring additional development.

See Appendix E for additional information on these criteria.

Appendix E

Examples of Criteria Used to Prioritize or Select Research Projects

This appendix provides excerpts from documents explaining project prioritization and selection criteria used by NSF, other agencies, and other countries. The majority of the text of this appendix consists of quoted material. The sources for these quotes are printed in italics, and explanatory material is enclosed in brackets.

NATIONAL SCIENCE FOUNDATION (1)

[Criteria for selecting projects for MREFC support, from *Rita Colwell's testimony before the House Committee on Science, Subcommittee on Research on September 6, 2001*]

- Intellectual merit
- Broader impacts
- Importance to science and engineering
- Cost-benefit and risk analysis
- Readiness to implement and manage
- Appropriateness for NSF
- Balance across fields and disciplines
- Synergy with other large facilities supported by NSF, other agencies, and other nations

NATIONAL SCIENCE FOUNDATION (2)

[Criteria for selection of projects for MREFC support, from *Facilities Management and Oversight Guide*, November 8, 2002]

- Exceptional S&E [Science and Engineering] opportunity to enable frontier research and education
  - Transformational regarding S&E impact
  - High priority within relevant S&E communities
  - Accessible to appropriately broad user community
  - Partnership possibilities exploited
  - Technical feasibility and risks addressed thoroughly
  - High state of readiness

NATIONAL SCIENCE FOUNDATION (3)

[Criteria for selection of projects for future MREFC support, from *Answers Provided by NSF to Questions from the House Science Committee Hearing on February 13, 2003*]

- Significance of the opportunity to enable frontier research and education
  - Degree of support within relevant S&E communities
  - Readiness of project, in terms of feasibility, engineering and cost-effectiveness, interagency and international partnerships, and management

NATIONAL SCIENCE FOUNDATION (4)

[Criteria for placing highly rated projects in priority order, from *Answers Provided by NSF to Questions from the House Science Committee Hearing on February 13, 2003*]

- How "transformative" is the project? Will it change the way research is conducted or change fundamental S&E concepts/research frontiers?
- How great are the benefits of the project? How many researchers, educators and students will it enable? Does it broadly serve many disciplines?
- How pressing is the need? Is there a window of opportunity? Are there interagency and international commitments that must be met?
- Additionally, the director must weigh the impact of a proposed facility on the balance between scientific fields, the importance of the project with respect to national priorities, and possible societal benefits.

NATIONAL SCIENCE BOARD

[From *NSB Resolution 02-191 adopted on November 21, 2002, Setting Priorities for Major Research Facilities*]

• Once construction for an approved and prioritized project commences, highest priority is given to moving that project forward through multiple years of construction in a cost-effective way, as determined by sound engineering and as long as progress is appropriate. It is most cost-effective to complete initiated projects in a timely way, rather than to commence new projects at the cost of stretching out in-progress construction.

• New candidate projects will be considered from the point of view of broadly serving the many disciplines supported by NSF.

• Multiple projects for a single discipline, or for closely related disciplines, will be ordered based on a judgment of the contribution that they will make toward the advancement of research in those related fields. Community judgment on this matter is considered.

• Projects will be authorized close to the time that funding requests are expected to be made.

• International and interagency commitments are considered in setting priorities among projects.

• The above are guidelines. Each facility consideration involves many complex issues. The Board will consider all relevant matters, and could deviate from these guidelines, given sound reasons to do so.

HOUSE SCIENCE COMMITTEE

[Priority setting, from *H.R. 4664, Section 14A.3*, the authorization bill for doubling NSF's budget]

• Scientific merit
• Broad societal need and probable impact
• Consideration of the results of formal prioritization efforts by the scientific community
• Readiness of plans for construction and operation
• The applicant's management and administrative capacity of large research facilities
• International and interagency commitments
• The order in which projects were approved by the [National Science] Board for inclusion in a future budget request

COSEPUP REPORT ON EVALUATING FEDERAL RESEARCH
(1999)

[Criteria for performance evaluation from *Evaluating Federal Research Programs: Research and the Government Performance and Results Act*, National Academies Press, 1999]

For basic research programs, agencies should measure quality, relevance, and leadership.
   • What is the quality of the research program—for example, how good is the proposed research work compared with other work being conducted in the field?
   • Is the proposed research focused on subjects most relevant to the agency mission?
   • Is the proposed research at the forefront of scientific and technological knowledge?

OECD FORUM ON MEGASCIENCE POLICY ISSUES

[Criteria for assigning priorities for major national research facilities, derived from *Australian Science and Technology Council Reports* by W. J. McG. Tegart, 1992]

Benefits to Science and Technology
Scientific objectives and their significance
   • Does the proposal develop an area of scientific or engineering research of great importance and which is at the leading edge of international research?
   • What are the key scientific questions that can be answered by having access to the proposed national facility?
   • Why are the answers to the questions significant for the national science and technology?
   • Will the proposed national facilities be made available to outside researchers subject to independent peer reviews?
Established need
   • Is the case for the proposed national facilities in terms of current national priorities?
   • Does the proposal involve a major source of expenditure on a piece or pieces of physical equipment of a scale that it could be developed incrementally or funded by an institution or consortium of institutions without serious disruption to other commitments of equal or higher priority?
   • Is there a community of outstanding scientists and technologists committed to the success of the proposed national facility?

Unique characteristics
- Are there characteristics of the proposed facility that are uniquely appropriate for the nation?

Degree of impact
- What impact will the proposed national facility have on fostering interdisciplinary research?
- Will the proposed national facility provide new opportunities for doctoral and post-doctoral training in research?
- Will the proposed facility contribute to public pride and the prestige of the nation's science and technology?

International characteristics
- Will the proposed national facility encourage international scientific collaboration by attracting researchers from overseas to spend time in the nation?
- Could the proposed national facility be located with advantage overseas in partnership with one or more other countries?
- Would the proposed national facility attract international partners?

Benefits to the Nation

Industry objectives and their significance
- Will the construction of the proposed national facility provide a technological stimulus to national industry?
- Will the proposed national facility provide unique services of benefit to national industry?
- Could the proposed national facility lead to better linkages between academic and research institutions and industry?
- Will the research output from the proposed national facility foster the development of new national enterprises?
- What contribution will the proposed national facility make to enhancing the skills base and training of national technology?

Social objectives and their significance
- Is the proposed national facility of high national priority for the advancement of knowledge, economic growth, health, welfare, or national security?
- Does the proposed national facility contribute to a better understanding and management of our environment?
- Will the proposed national facility lead an improved understanding and appreciation by the national community of the accomplishments of science and technology?

International standing
- Will the proposed national facility project and enhance the nation's image as a technologically advanced nation?
- Will the nation's position in international negotiations be strengthened as a result of the proposed national facility?

NIH STRATEGY FOR SETTING RESEARCH PRIORITIES

[From *NIH Director Harold Varmus's testimony before the House Appropriations Subcommittee on June 10, 1997*]

- Exploit recent discoveries, such as the isolation of new genes for human diseases
- Encourage studies of diseases that have been relatively neglected, poorly controlled, or recently made more accessible to scientific study
- Strengthen research technologies, such as computer science, imaging devices, neuroscience, or gene mapping, applicable to a broad range of disciplines and diseases.

THE NATIONAL RESEARCH PRIORITIES TASKFORCE

[Discussion excerpted from a report commissioned by the Australian government, *Developing National Research Priorities*, 2002]

Ultimately, for national priorities to be worthwhile they must have three important characteristics:

- An increased research effort must be capable of delivering a measurable and significant positive impact on the objective underlying the priority
- Australia must be able to build the capacity needed to achieve that impact
- Australia must be able to capture the benefits of that research (either through its commercialization or application).

Based on this, the government has developed three criteria for assessing nominated priorities. These criteria will be used by the expert advisory committee in developing the short-list of priorities for government consideration.

- The scope for Commonwealth Government investment in research to have a measurable and significant positive impact, by:
  - ◦ achieving a 'critical mass' through specific support and/or coordination and collaboration at the national level
  - ◦ addressing uniquely Australian needs and challenges arising from our geography, climate, bioresources, economy, way of life and/or culture.
  - ◦ The scope for Australia to build quickly the capacity to achieve that impact, taking into account:
  - ◦ existing expertise, experience and technological capacities or whether such capacities can be reasonably acquired

- ○ the availability, quality and scale of necessary research infrastructure
- ○ the strategic research priorities of other nations and the potential benefits of international collaborations
- ○ the overall magnitude of the investment required to make an impact.
- The scope for Australia to capture the benefits of the research, through the potential of the research to:
  - ○ achieve commercially or socially relevant outcomes over the cycle of the priorities regime
  - ○ enhance significantly Australia's overall innovation capacity by the broadening of the knowledge base, and fostering acquisition of skills and understanding of emerging 'hot' sciences.

DECISION CRITERIA FOR EVALUATING AGRICULTURAL RESEARCH PRIORITIES

[Report from *Northeastern Regional Association of State Agricultural Experiment Station Directors*, 1999]

Selecting the topics for future research investments in agricultural science requires a rationale that is defensible, and a process that is, for the public sector, completely transparent. This document sets out a system of points-to-consider that could make up a new approach to setting criteria for decision making on resource allocations. The fundamental goal for setting agreed-to-decision criteria for allocating public sector agricultural research should be to gain the largest possible returns on research investment. This then requires consideration of four factors:

- There must be a congruence between the dimensions of the intended topic and the allocation of research resources. Larger impacts can be expected by investing in topics that already have a large base in agriculture, forestry or rural development (e.g., wheat, hardwoods, community services), rather than trying to start from a smaller base (e.g., edible amaranth) or a regionally distributed environmental issue rather than local or state. This requires that some congruence analysis be done.
- The allocation of resources needs to directly reflect the needs of the intended stakeholders and customers. This requires us to be listening to our customers.
- Judgments are needed on what is feasible to accomplish through agricultural research, and this needs to be grounded in the best possible science. This in turn mandates some evaluation of the scientific potential of proposed research approaches by knowledgeable scientists.

• Projections of expected benefits are needed to permit informed choices of alternative allocations. These must be done with a set of assumptions that are understood by the participants and the claimants to the system, and are broadly agreed upon. Ex ante estimations need to go beyond economic consequences to the non-economic benefits and consequences of technologies (i.e., social, environmental). Assigning premium or discount coefficients to economic projections can do this.

SYSTEMATIC RESEARCH PRIORITY CRITERIA

[From *UK Systematics Forum on Priorities in Systematic (Biology) Research and Training*, 1995]

• *Scientific excellence.* The scientific excellence of the proposed study, the individual or team who will undertake the work and the collections available for study.
• *Relevance.* The relevance of the study to a significant and clearly identified user community. The relevance may be scientific, cultural, historical or economic and will vary in relation to the differing interests of different audiences or user communities.
• *Enhancing scientific coverage.* Filling gaps in knowledge about a group of organisms and making information available. Gaps in knowledge are not a sufficient criterion for establishing high priority, unless criterion 2 also applies and the results are of immediate relevance.
• *Scale of relevance.* Levels of priority may differ at different geographical levels (global, regional, national or local).
• *Urgency.* The level of priority given to a proposed study may reflect degree of urgency of the work determined in relation to the endemicity of the [proposing] group.
• *Feasibility.* The scientific feasibility of completing the proposed task in the timescale and with the resources proposed.

UK DEPARTMENT OF HEALTH

[*Prioritization of the Policy Research Programme in collaboration with the R&D Directorate*]

• Ministerial priority and relevance to the goals, aims and objectives of the Department of Health
• Size and importance of the problem to be addressed in terms of actual or potential burden of disease and social condition
• Well-defined plans for introducing research results into current policy activity or the formulation of future policy

- Timeliness
- Feasibility of research
- Likely return on the investment in research
- Appropriateness and availability of other research budgets

GERMAN SCIENCE COUNCIL

[From the German Wissenschaftsrat's *Large-scale Facilities for Basic Scientific Research, Working Group Report*, 2003]

Theses on the significance of large-scale facilities for basic scientific research:
- Scientific and technical developments for industry and society often start with the findings of basic research.
- The opening-up or development of totally new areas of research is closely related to the availability of specific new facilities.
- Large-scale facilities should stem from a broad initiative of scientific users with equal rights.
- It is important to make very sure that future generations of scientists receive adequate, forward-looking training using the large-scale facilities.
- With large-scale facilities on the scale with which we are concerned here, there must be long-term scientific visions, and the prerequisites for technical innovations must be given.
- Including foreign partners in project preparation at an early stage, letting them participate in decision making and be responsible for their own scientific or technological contributions are prerequisites.
- In order to avoid the duplication of research infrastructures, which would be detrimental to the effective capacity usage of large-scale facilities, there should be no comparable rival projects at the national or European level that are already in the realization phase.
- Large-scale facilities must be centrally incorporated into the strategic planning and research program of the institution(s) in charge and must be a core element of the spectrum of tasks.
- In the field of apparatus and equipment, which includes large-scale facilities, cooperation between places of further education and major research establishments should be a matter of course. Access to the facilities must be kept open for scientists from universities in a systematic and pragmatic way, so as to allow them to carry out research projects.

Criteria and specific considerations:
Expert Panel
- The probability of fundamentally new insights or the possibilities of decisive scientific advances which could only be achieved with the large-scale facility.

- The large-scale facility's technical feasibility and the degree of technical innovation.
- The scientific and technical competence of the institutions involved.
- The already existing or anticipated acceptance of the (potential) users from immediately relevant and from neighboring areas of expertise.
- The fulfillment of various objectives of importance for research (transfer, international perspectives, promoting young scientists).

Policy Assessment Panel
- Scientific potential of the research program.
- Fulfillment of science and technology policy goals as formulated in the theses.
- Degree of maturity of the technical concept and, linked to it, the possible timeframe for implementing the individual facilities.
- The context of further national and international scientific development of the research fields they belong to and assess their interaction with other disciplines.

[See Appendix D for additional information on this process.]

UK OFFICE OF SCIENCE AND TECHNOLOGY

[From *UK Office of Science and Technology's Large Facilities Strategic Roadmap Report*, 2003]

[The UK uses a "roadmap" portfolio approach to select projects for solicitation and a Gateway screening process that requires projects to meet certain conditions before moving forward. The concept behind the Gateway process is that all large capital investments should be managed as discrete projects and should be subject to review and independent scrutiny at all key stages in their life cycle. The Gateway Process is based on well-proven techniques that lead to more effective delivery of benefits together with more predictable costs and outcomes. The process considers the project at critical points in its development. These critical points are identified as Gateways. There are six Gateways during the lifecycle of a project, four before contract award and two looking at service implementation and confirmation of the operational benefits. The Process emphasizes early review for maximum added value.]

The four-stage Gateway process:
- *Strategic assessment*. Initial assessment of the strength of the science requirement for potential projects. Scientific peer review must be a key element of this first review.

- ◦ Importance [depth] of science knowledge to be delivered by project.
- ◦ Breadth of science knowledge that will benefit from investment.
- ◦ Match with international positioning of UK science.
- ◦ Strength of opportunity for training [links to number of users].
- ◦ Contribution to/from UK technology/industry base.
- ◦ Opportunity for spin-off and exploitation.
- *Business justification.* Justification and robustness of the business case.
- *Procurement strategy.* Confirm procurement strategy, project plan, etc.
- *Readiness for service.* Confirm contract decisions and let contracts.

[See Appendix D for additional information on this process.]

Appendix F

NSF Background Materials

W ide assortments of materials were made available to the committee in the course of its work. The committee includes here excerpts from several of them that highlight the current set of principles and practices at NSF.

HEARING ON NATIONAL SCIENCE FOUNDATION MANAGEMENT OF LARGE FACILITIES[1]

Statement of Dr. Anita Jones, vice chair, National Science Board
Before the House Committee on Science Subcommittee on Research,
September 6, 2001

Chairman Smith, Ranking Minority Member Johnson, members of the Committee, I appreciate the opportunity to testify today. My name is Anita Jones. I am Vice Chair of the National Science Board and Chair of the Board's Committee on Strategy and Budget. I am also the Quarles Professor of Engineering and Applied Science at the University of Virginia. From 1993 to 1997 I served as Director of Defense Research and Engineering at the U.S. Department of Defense. In that position I was responsible for the science and technology program of the Department of

---

[1]The text of these remarks was obtained from http://www.nsf.gov/od/lpa/congress/107/jones_facilities90601.htm.

Defense, including the Defense Advanced Research Projects Agency and oversight of the Department's laboratories.

On behalf of the National Science Board, I thank the Committee for its long-term support for science and engineering research and education activities, which have contributed so substantially to our Nation's well being.

The National Science Board has two statutory roles: to serve as the governing board of the National Science Foundation, and to advise the Congress and the President on national policy issues for science and engineering research and education.

Today, my comments will focus on the Board's role as governing board of the Foundation, specifically on our oversight and approval of the Foundation's support for large-scale research facilities.

First, I would like to emphasize that the Foundation has an excellent record—spanning 50 years—of supporting such facilities, in terms of both the quality of their research and their management. Today, NSF invests over $1 billion annually in facilities and other infrastructure projects. With the exception of U.S. research facilities in the Antarctic, which are directly operated by the National Science Foundation, NSF typically makes awards to other organizations for the construction and operation of facilities.

The following are examples of major facilities:

• The Large Hadron Collider is a superconducting particle accelerator. Its purpose is to help scientists advance the fundamental understanding of matter. The Collider's construction and operations are funded through an international collaboration.

• The Laser Interferometer Gravitational-Wave Observatory, or LIGO, will allow physicists and engineers to collaborate to test the dynamic features of Einstein's theory of gravity and to study the properties of intense gravitational fields.

• The National Astronomy Center in Arecibo, Puerto Rico, supports observations in radio and radar astronomy and atmospheric sciences.

• Research facilities at the South Pole Station, currently under renovation, support a variety of diverse but important research activities that can only be conducted in the unique cold and pristine environment at the South Pole.

• The Ocean Drilling Program, involving 20 countries, supports research in areas including deep ocean structures, hydrology and geochemical cycles.

These five examples are all major research facilities. For the most part, they are the research instruments that make possible research advances that can be accomplished in no other way. They are all large; each one opens new research frontiers that could not be entered without these tools. They are complex; each one involves challenging engineering tasks in its design, construction and operation. Hence, each is very costly. As I have illustrated, the U.S. frequently teams with international collaborators, not just to assure that the best research is pursued, but also to help make the facilities more affordable. And, very importantly, each facility has a very broad base of researchers who are the users; they frequently come from multiple disciplines.

Many scientific fields are on the edge of exciting discoveries that require such facilities. I anticipate that in the 21st century, the need for such large, complex research facilities will grow. In recent years, the Foundation's portfolio of facilities has grown and diversified to include distributed projects and complex multidisciplinary projects like terascale computing systems and ocean observatories that challenge traditional management and oversight approaches.

National Science Board's Oversight of Large Facility Projects

The National Science Board plays a critical role in the oversight and approval of large NSF-supported facilities. The NSB is well constituted to exercise its oversight and approval responsibilities. Members of the Board include executives from industry and presidents of universities, individuals who have extensive experience in managing large, cutting-edge research facilities and instrumentation. Of course, the Board includes members who have used such facilities.

The Board conducts two activities that focus on the approval and oversight of facilities. They are the approval of large awards and the approval of candidates for the Major Research Equipment account. Typically, the Board hears briefings from NSF management at almost every NSB meeting on the subject of large facilities—existing and candidate.

NSB Approval for Major Awards

The Board approves all major projects, including facilities, whose costs exceed one percent of the budget of the sponsoring directorate or office. The Board also approves new major programs whose budget exceeds three percent of the budget of the sponsoring directorate or office.

The Board's Committee on Programs and Plans (CPP) reviews large projects at various stages of their development and makes recommendations to the full Board for the initiation of new awards and programs. In addition, the CPP reviews projects for adherence to the NSB approved criteria for merit review and the Board's policy regarding the competition, recompetition and renewal of NSF awards. Throughout the implementation phase of a project, the CPP reviews its progress and informs the Board of its status and any issues that arise.

The Board's Committee on Audit and Oversight (A&O) reviews specific financial and business management issues raised by the Inspector General and by Foundation management. Like the CPP, A&O informs the Board of any issues that arise.

The Major Research Equipment Account

The Major Research Equipment (MRE) account is an agency-wide capital asset account used to fund major science and engineering infrastructure projects that cost far more than one program's budget could support. The costs of MRE facilities range from several tens to hundreds of millions of dollars. The Board sees these projects multiple times over their lifetime. The Board takes two kinds of actions. First, it authorizes a candidate project for possible inclusion by the Foundation in a future budget. Later, the Board approves specific funding for an organization or consortium to design and construct the facility.

Let me briefly outline how the Board oversees MRE projects. The Director selects candidates for the NSB to review during one of its five meetings throughout the year. The Board receives, for approval, candidate projects that may be included in a future budget request, subject to availability of funding. Board authorization signifies that the projects are meritorious and that planning is sufficiently advanced to justify funding. In giving its approval, the Board considers the intellectual merit, societal impacts of the projects, their importance to science and engineering, balance across disciplines, readiness to be implemented, and cost-benefit and risk analyses.

The Board authorizes the MRE projects for possible inclusion in future budgets, but does not rank-order them to preserve the Foundation's flexibility in a given budget year. We believe that all projects authorized by the Board are of unquestioned excellence and worthy of Foundation support. When the Board approves the Foundation's budget submission to the Office of Management and Budget, it reaffirms its support for any MRE projects included for funding.

After NSF has run a competition and detailed plans are in place for design, construction and operations, the project comes back to the Board for the award of funding at a specific level. Board oversight of MRE projects continues after an award is made. The Board's CPP reviews a project's progress at the midpoint of construction and whenever significant issues arise. If it appears that a project will exceed the Board's approved dollar amount by over 20 percent or $10 million, whichever is less, the Director must return to the Board to request approval for a higher level of funding. CPP also receives periodic reports on the status of all major projects.

NSF Large Facility Projects Management and Oversight Plan

As part of its oversight responsibilities, the Board began a dialogue with Foundation management about large facility issues more than a year and a half ago. Together, we have been discussing the improvement of the process for identifying candidate projects and for Foundation oversight of the management of construction and operation of such facilities.

The Foundation has created a new Large Facility Projects Management and Oversight Plan. That Plan, which was requested by the Administration, does two things. It incorporates and builds upon an existing facility management process. In addition, it strengthens financial oversight.

I would like to comment briefly on the Board's participation in the development of the Plan. The full Board received a draft Plan for comment this summer. At the August 8-9 Board meeting, our Committee on Programs and Plans, which has the responsibility to review major projects and facilities, received a briefing on the Plan from the Deputy Director. Members of our Audit and Oversight Committee participated in those discussions. Board members were pleased with the direction, framework, and elements set forth in the briefing and encouraged Foundation management to proceed with the Plan's development. The Board will continue to assess the Foundation's progress in refining and implementing the Plan.

The Plan institutionalizes and builds on long-standing management practices.

- It codifies sound practices already in use, augments the existing MRE process, and documents principles of management.
- It ensures that project management will stay with the scientists and engineers, from planning through operation. The overall NSF Program Manager for a particular facility project is an individual in one of the research directorates of the Foundation.
- It strengthens Foundation oversight of financial and business functions. This requires organizational and managerial changes within the Foundation. In particular, it calls for the creation of a Deputy for Large Facility Projects who reports directly to the Chief Financial Officer and is responsible for "developing, implementing and managing, with NSF-wide input and concurrence, management oversight policies, guidelines and procedures."

In summary, Mr. Chairman and members of the Committee:

• The National Science Board supports the general direction laid out in NSF's Large Facility Projects Management and Oversight Plan. The Board will assess the Foundation's progress in refining and implementing the elements of the Plan, particularly to ensure the integrity of the evaluation and oversight of the financial and business aspects of the facility project throughout its life.
• The implementation of the Plan will ensure that the Foundation, with Board oversight, has the policies and organization required for sound management of unique, complex, world-class research facilities.

Thank you for the opportunity to present these remarks. I would be glad to answer any questions you may have.

EXCERPTS FROM THE NSF FY04 BUDGET REQUEST[2]

MAJOR RESEARCH EQUIPMENT AND FACILITIES CONSTRUCTION $202,330,000

The FY 2004 Budget Request for Major Research Equipment and Facilities Construction (MREFC) is $202.33 million, an increase of $76.05 million, or 60.2 percent, above the FY 2003 Request of $126.28 million.

MREFC Funding
(Dollars in Millions)

	FY 2002 Actual	FY 2003 Request	FY 2004 Request	Change Amount	Percent
Major Research Equipment & Facilities Construction	$115.35	$126.28	$202.33	$76.05	60.2%

The MREFC Account supports the implementation of major research facilities and equipment that provide unique capabilities at the frontiers of science and engineering. Implementation projects supported by this account are intended to extend the boundaries of technology and open new avenues for discovery for the science and engineering community. Initial concept and development, and follow on operations and maintenance costs of the facilities are provided through the Research and Related Activities (R&RA) Account.

There can be no doubt that a modern and effective research infrastructure is critical to maintaining U.S. leadership in science and engineering (S&E). The future success of entire fields of research depend upon their access to new generations of powerful research tools. Increasingly, these tools are large and complex, and have a significant information technology component.

Among Federal agencies, NSF plays a major role in providing the academic (non-medical) research community with access to forefront instrumentation and facilities. In recent years, NSF has received an increased number of requests for major research facilities and equipment from the S&E community. Many of these requests have been rated outstanding by research peers, program staff, management and policy officials, and the National Science Board. NSF's request for the MREFC Account fully funds the ongoing projects and the remaining three projects approved for funding by the National Science Board, but not yet funded, and positions the agency to meet the future needs and opportunities of the research community.

Once a project has been submitted for MREFC funding, it must undergo a multi-phase review and approval process. The process begins with a review by the MREFC Panel, which makes recommendations to the NSF Director with attention to criteria such as scientific merit, importance, readiness and cost-benefit. The Director then selects candidates for National Science Board (NSB) consideration. The NSB then approves, or not, projects for inclusion in future budget requests and establishes priorities. The Director selects from the group of NSB-approved projects those appropriate for inclusion in a budget request to OMB, and after discussion with OMB, to the Congress.

In order for a project to be considered for MREFC funding, NSF requires that it represent an exceptional opportunity that enables research and education. In addition, the project should be transformative in nature, in that it should have the potential to shift the paradigm in scientific understanding and/or infrastructure technology. NSF believes that all the projects included in this Budget Request meet these criteria.

As a general framework for priority-setting, NSF assigned priority to projects based on the following criteria:

"   First Priority: Ongoing Projects – Projects where outyear funding for the full project has already been included in a Budget Request to Congress, and projects that have received initial funding for startup operations.

---

[2]The text of the NSF FY2004 budget request to Congress for the MREFC account as obtained from http://www.nsf.gov/bfa/bud/fy2004/pdf/fy2004_16.pdf. Reproduced here are pages 2-3.

Major Research Equipment and Facilities Construction

"　　Second Priority: NSB-Approved New Starts – New projects that have received NSB approval for inclusion in a budget request but which have not yet been included in a budget request or received funding.

NSF believes that the highest priority within the MREFC Account must be the current projects. To that end, highest priority in FY 2004 is to continue to request funding for:

"　　Atacama Large Millimeter Array Construction ($50.84 million);
"　　EarthScope: USArray, Plate Boundary Observatory and San Andreas Fault Observatory at Depth ($45.0 million);
"　　The High Performance Instrumented Airborne Platform for Environmental Research ($23.53 million);
"　　The IceCube Neutrino Observatory ($60.0 million);
"　　The George E. Brown Network for Earthquake Engineering Simulation ($8.0 million);
"　　The National Ecological Observatory Network ($12.0 million); and
"　　South Pole Station ($960,000).

In addition, three new starts are requested in FY 2005 and FY 2006. In priority order, these are: Scientific Ocean Drilling in FY 2005; Rare Symmetry Violating Processes in FY 2006; and Ocean Observatories in FY 2006.

NSF Funding for MREFC Projects, FY 2002 through FY 2008[1]
(Dollars in Millions)

	FY 2002[2] Actual	FY 2003 Request	FY 2004 Request	FY 2005 Request	FY 2006 Request	FY 2007 Request	FY 2008 Request
ONGOING PROJECTS							
ALMA Construction	12.50	30.00	50.84	49.67	48.84	47.89	46.49
EarthScope: USArray, SAFOD, PBO		35.00	45.00	54.26	40.00	23.00	
High-performance Instrumented Airborne Platform							
for Environmental Research	35.00		25.53				
IceCube Neutrino Observatory	10.12		60.00	33.40	34.30	35.30	36.30
Polar Aircraft Upgrades	0.89						
Large Hadron Collider	16.90	9.72					
Network for Earthquake Engineering Simulation	24.40	13.56	8.00				
National Ecological Observatories Network[3]		12.00	12.00	16.00	20.00	20.00	20.00
South Pole Station	15.55	6.00	0.96				
Terascale Computing Systems		20.00					
NEW STARTS							
Scientific Ocean Drilling				76.85	23.00		
Rare Symmetry Violating Processes					30.00	42.66	44.00
Ocean Observatories					24.76	40.33	72.46
Totals	$115.35	$126.28	$202.33	$230.18	$220.90	$209.18	$219.25

[1]Does not include funding provided for early concept and development or follow-on operations and maintenance. These funds are provided through the R&RA Account and are discussed in the following individual Activity narratives and the Tools section.
[2]FY 2002 Actuals include $16.44 million in carryover from prior year appropriations for the South Pole Station Modernization Project, the South Pole Station Safety and Environment Project, and the Polar Aircraft Upgrades. $39.88 million appropriated in FY 2002 is carried over into FY 2003 for the IceCube Neutrino Observatory and Terascale Computing Systems. This FY 2002 carryover will be reflected in the Current Plan following an FY 2003 appropriation.
[3]FY 2006-08 implementation funding for NEON will be contingent upon the outcome of the feasibility study of the NEON project and the successful review of the prototype NEON sites.
[4]FY 2002 funding for Terascale was carried over into FY 2003 due to the NSB meeting schedule. The award was approved in October, 2002 and the funds have been obligated.

EXCERPTS FROM ANSWERS PROVIDED BY NSF TO QUESTIONS FROM THE HOUSE SCIENCE COMMITTEE HEARING, 13 FEBRUARY 2003[3]

Question 1: The President's fiscal year 2004 budget request represents a significant improvement over prior year requests with regard to directorate breakdown, full life-cycle costs, and prioritization of the "new start" projects in the Major Research and Facilities Construction (MREFC) account However, the budget does not include the criteria used to rank these projects and the relative value these criteria were given in establishing the prioritized list. This information is required to be annually submitted to Congress before any funds can be obligated from the MREFC account. Please provide us with this information. Also, please clarify how new projects will be reviewed and ranked, how they will be incorporated into the existing prioritized list, and when and how the revised list will be transmitted to Congress.

Answer: In FY 2004 budget request, $202.33 million in funding is requested for seven ongoing MREFC projects. No funds were requested for new start projects. This is consistent with current NSB policy, which requires that NSF give first priority to projects that have been started but not completed. The FY 2004 budget request identified three new starts for initiation in FY 2005 and FY 2006. In priority order, these are:

- Scientific Ocean Drilling in FY 2005;
- Rare Symmetry Violating Processes in FY 2006; and
- Ocean Observatories in FY 2006.

This specific case (i.e., the process that NSF used to prioritize these three projects) must first be viewed within the broader context of how NSF identifies, reviews, selects and prioritizes large facility projects.

The Broader Context

In identifying new facility construction projects, the S&E community, in consultation with NSF, develops ideas, considers alternatives, explores partnerships, and develops cost and timeline estimates. By the time a proposal is submitted to NSF, these issues have been thoroughly examined.

---

[3]The text for this written transmittal from NSF to Congress was obtained in printed form from the Deputy Director's office at NSF.

Upon receipt by NSF, large facility proposals are first subjected to rigorous external peer review, focusing on the criteria of intellectual merit and the broad (probable) impacts of the project. Only the highest rated proposals—i.e., those that are rated outstanding on both criteria—survive this process and are recommended to an MREFC Panel comprised of the Assistant Directors and Office heads, serving as stewards for their fields and chosen for their breadth of understanding, and chaired by the Deputy Director.

The MREFC Panel uses a two-stage process. First, it selects the new start projects it will recommend to the Director for future NSF support, based on a discussion of the merits of the science within the context of all sciences that NS supports. Second, it places these recommended new start projects in priority order.

In selecting projects for future support, the Panel considers the following criteria:

- Significance of the opportunity to enable frontier research and education.
- Degree of support within the relevant S&E communities.
- Readiness of project, in terms of feasibility, engineering cost-effectiveness, interagency and international partnerships, and management.

Using these criteria, projects that are not highly rated are returned to the initiating directorates, and may be reconsidered at a future time. The Panel then places highly rated projects in priority order. This process is conducted in consultation with the NSF Director. The MREFC Panel and the Director use the following criteria to determine the priority order of the projects:

- How "transformative" is the project? Will it change the way research is conducted or change fundamental S&E concepts/research frontiers?
- How great are the benefits of the project? How: many researchers, educators and students will it enable? Does it broadly serve many disciplines?
- How pressing is the need? Is there a window of opportunity? Are there interagency and international commitments that must be met?

These criteria are not assigned relative weights, because each project has its own unique attributes and circumstances. For example, timeliness may be crucial for one project and relatively unimportant for another. Additionally, the Director must weigh the impact of a proposed facility on the balance between scientific fields, the importance of the project with respect to national priorities, and possible societal benefits.

In August the Director presents the MREFC priorities, including a discussion of the rationale for the priority order to the NSB, as part of the budget process. The NSB reviews the list and either approves or argues the order of priority. As part of its budget submission, NSF presents this rank-ordered list of projects to OMB. Finally, NSF submits a prioritized list of projects to Congress as part of its budget submission.

The Specific Case

The three new start projects cited in the FY 2004 budget request are considered highly meritorious by the S&E community, the NSF and the NSB.

The Scientific Ocean Drilling (SOD) Project was ranked as the highest priority because delaying initiation of the project until FY 2006 would greatly impact this existing community of researchers, and because of the significant level of complementary international effort and planning that is already underway. This project will charter and modify a drill ship which will work in a new scientific program (Integrated Ocean Drilling Program [IODP], in concert, and complementary to, a deep drilling vessel to be constructed and operated by Japan. Some of the drilling to be done from the SOD vessel will be used to guide and plan drilling from the Japanese vessel, which is scheduled to begin operations in 2007. Additional international members who help finance our existing ocean drilling program are prepared to join the new program, but will have trouble maintaining and committing their financial contribution if drilling from the SOD vessel is delayed until 2007. At present, the Japanese vessel has been constructed and is undergoing outfitting. If the U.S. does not meet its commitment, there will be no conventional drill ship capabilities for use in the IODP, and critical studies of climate change and the ocean biosphere will be jeopardized.

The two remaining new start projects, RSVP and Ocean Observatories, were felt to be of equal value, but for different reasons. RSVP ranked second, primarily for reasons of balance across scientific fields. RSVP is very well designed, well reviewed, and addresses important scientific questions that have the potential to transform our basic understanding of the universe. There are, as with SOD, performance and cost risks associated with delays. The host laboratory, Brookhaven National Laboratory, would be forced to lay off key staff and then rehire and/or replace them following an extensive shut-down of beams planned for use by RSVP; and, the international collaborators may have difficulty maintaining (as SOD will) the large financial contributions currently committed to RSVP, on order of $10 million (US). Nevertheless, these considerations do not outweigh the funding and stewardship issues represented in SOD. If initiated in FY 2006, RSVP can still be implemented successfully and make major contributions to science.

The Ocean Observatories project will enable a large group of researchers to perform ocean science in new ways. It was ranked third among the new start projects because it is not as urgent as SOD or RSVP, and again, for reasons of balance across scientific fields.

Appendix G

Executive Summary of COSEPUP Report

MAJOR AWARD DECISIONMAKING AT THE NATIONAL SCIENCE FOUNDATION

Panel on NSF Decisionmaking for Major Awards

Committee on Science, Engineering, and Public Policy

National Academy of Sciences
National Academy of Engineering
Institute of Medicine

NATIONAL ACADEMY PRESS
Washington, D.C. 1994

This report assesses and makes recommendations to strengthen the merit review system used by the National Science Foundation (NSF) to make major awards to support important research facilities, centers, and other large-scale research-related activities. The purpose of the recommendations is to ensure that the most meritorious projects are chosen for support, that the selection process is fair in practice and perception, and that the results in each case are clearly and publicly explained. In this way, the effectiveness and accountability of the major award process will be increased, and the confidence of the research community, Congress, and the public in the system will be enhanced.

The United States has built the most successful research system in the world. The use of peer review to identify the best ideas for support has been a major ingredient in this success. Peer review-based procedures such as those in use at NSF, the National Institutes of Health, and other federal research agencies remain the best procedures known for ensuring the technical excellence of research projects that receive public support. Today, the nation is facing serious international economic competition, which extends to scientific and engineering research. To maintain our world class research enterprise, we will have to be more careful than ever to choose wisely the projects that receive support. The difference between an excellent proposal and one that is merely above average is critical in this effort. The merit review system must be maintained and strengthened to perform the function of choosing the best research for public support.

BACKGROUND

During the past decade, NSF has established Engineering Research Centers, Supercomputer Centers, Science and Technology Centers, and other large research centers and facilities. A few awards were controversial, and called into question NSF policies and procedures for making large award decisions. Some of those involving the location of one-of-a-kind national facilities have generated the sharpest questions about selection procedures. Decisions by the National Science Board (NSB) and the NSF to devote substantial resources to some new center programs and very expensive facilities have also raised questions about the adequacy of their planning procedures. The congressional conference report on FY 1991 appropriations for NSF requested a National Academy of Sciences (NAS) study of the criteria weighed in making major awards and an assessment of the roles of outside experts and agency staff in the merit review decisionmaking process at NSF. The NAS agreed to undertake the project because of the importance of merit review for making major research awards. The study was assigned to the Committee on Science, Engineering, and Public Policy (COSEPUP), which is chartered by the NAS, the

National Academy of Engineering, and the Institute of Medicine to address important questions that cut across all areas of science and engineering.

COSEPUP, with the approval of the president of the NAS, appointed a panel with a broad range of expertise to carry out the study (Appendix A). The panel studied NSF's policies and procedures governing major awards, defined as those awards for research and related activities that are subject to approval by the NSB because of their cost. Members of the panel consulted with past NSF directors, current officials, and NSB members, and examined in detail 10 case studies of major awards for research centers and facilities (listed in Appendix E). The NSB reviews between 30 and 50 decisions a year on major awards, which account for about 30 percent of NSF's Research and Related Activities budget of $2.0 billion in FY 1994 (recent awards are listed in Appendix C).

The panel carefully examined the cycles that each major award goes through. These included the processes leading to the initial decision to announce a major project; the planning and implementation of the merit review process; the decisionmaking leading to the award; and subsequent decisions to renew, recompete, or terminate a project at appropriate intervals. The panel focused on the roles of expert peer reviewers, staff, outside advisory groups, and NSB in the merit review process, and on the public explanation of the process, and its outcomes.

In addition to examining NSF policies and procedures, and the organization and resources it has to carry them out, the panel focused on the role and capacity of NSB in discharging its legal authority for design of the review process and for approving each major award. At each stage, NSB has an opportunity to approve, cancel, or postpone further action.

FINDINGS AND CONCLUSIONS

The panel concluded that merit review has generally served well to ensure fairness, effectiveness, and efficiency in decisionmaking on research projects over the years, but for major awards the system needs some changes to accommodate evolving conditions and special features of costly large-scale, long-term projects. NSF has successfully made many highly visible and important awards with relatively few controversies. The merit review system has been the major reason for the high quality of the activities selected for support, and it has served to discourage the use of inappropriate or parochial considerations in the allocation of NSF's research finding. Merit review is not perfect, but no clearly superior method of selecting research and research-related projects for support has been discovered after many years of experience here and abroad.

Although controversial decisions have been relatively rare, they have revealed problems in NSB and NSF policies and procedures that could be avoided. When such problems occur or are believed to occur, they undermine the confidence in the merit review system of the research community, research institutions that compete or hope to compete for major awards in a fair process, and Congress. So far, the success of the merit review system has helped insulate NSF and NSB decisionmaking on major awards from congressional intervention. If confidence in the system is not maintained, the temptation for research institutions to try to have Congress preempt NSF decision making will increase, and to the extent that legislative involvement replaces merit review with political considerations in project selection, the quality of the nation's research system may be negatively affected.

The panel recommends a number of changes to strengthen or improve the planning, review and selection, and subsequent renewal of major awards. Detailed recommendations are contained in various chapters of the report, but the key points follow:

Clear Rules of the Game

The "rules of the game" (i.e., the criteria, procedures, and roles of participants in the merit review process) must be absolutely clear in advance.

In some cases, the criteria or requirements needed to meet them have not been clear or were seemingly redefined during the review process. Although too much detail in specifying criteria might limit the flexibility to respond to innovative proposals, we concluded that to increase procedural fairness, NSB and NSF should be more precise about the criteria and review process to be used. In particular, the primary technical criteria as distinct from other criteria to be considered in the merit review process should be identified in advance in each case.

The panel recommends stronger planning efforts that would help contribute to clearer criteria (Recommendation 1). The panel also recommends that NSF concentrate more effort in designing a better-understood review process for each major award (Recommendation 8).

Primacy of Technical Merit

Technical merit must be the primary consideration in making awards.

The panel strongly supports the primacy of technical merit in the selection of major projects (Recommendation 3), and it endorses the use of a two-phase review process that would clearly indicate the ranking of projects on technical merit before other merit factors are considered (see next section).

Technical merit must be paramount to maximize the likelihood that the project will achieve its substantive research goals. Other criteria of merit should also be given due consideration in selecting the overall winner or winners, but any project receiving an award should rank among the very highest in technical quality. That should be made clear to all reviewers and decisionmakers, along with a sense of the nature and relative priority of each of the criteria.

NSF and NSB must be clearer in each case about the relative priority of the various criteria used, especially of the technical relative to the nontechnical criteria. Otherwise, the weightings of criteria are implicit and can shift continually at the discretion of individual reviewers and program staff.

Appropriate Roles of Peer Reviewers and Staff

The review process must be structured so that the roles of peer reviewers and staff in evaluating and recommending proposals are clearly understood, and trade-offs among technical and other criteria are clearly explained, at each subsequent level of decision making.

Currently, the summary rating and ranking of proposals by staff at various decision points does not always distinguish peer review from staff judgments. Although staff should make their best case for a recommended decision, the NSF director and the NSB should always know the results of the peer review.

The two-phase review process, properly documented, would make it easier to implement this objective (Recommendation 6). This two-phase process would facilitate the preparation of a summary document that explains the rationale for the decision, including the treatment of peer review results and the trade-offs made between technical and nontechnical criteria in reaching the final decision (see next section).

Public Documentation of Decision Making

There should be a public document explaining the results of the review and the rationale for the final decision by the NSB.

NSB minutes rarely record the basis for a major award decision, and no public document of explanation for the final decision is prepared or disseminated. The lack of such documentation leads to public confusion and controversy that could be avoided.

The panel recommends a short, carefully prepared memorandum that summarizes the results of each stage of the merit review process and outlines the rationale for choosing a winning proposal (Recommendation 9). Such memoranda would increase public understanding of major award decisions and therefore enhance public confidence in the system that produces them.

More Stringent Setting of Priorities

Decisions to solicit proposals for very large major awards should take into account their impact on NSF's overall program as well as on the particular research field involved, and they should be contingent on the realization of expected funds and technological progress.

Careful front-end planning, combined with broad consultation with affected research communities and constant evaluation of priorities at each decision point, must be a part of the process of soliciting and reviewing proposals for a very large major award. Solicitations for awards that have serious long-range budget implications must be based on a broader range of considerations than in the past. The priority within a given field should be clearly established and compared with the overall priorities of NSF across fields. After initial approval of a large project, contingency plans for possible unfavorable program or budget developments should be made for each project and updated annually. The potential impact on NSF priorities if there are unrealized budgetary expectations or unexpected technological problems or opportunities should be carefully reviewed at each decision point. In this way, NSF and NSB would avoid letting a series of small decisions in the development of a major project result in a project that no longer matches the agency's overall program priorities or budget.

The panel calls for stronger planning efforts, including contingency plans for lower funding levels than expected (Recommendation 1), based in part on a broader range of input from research communities affected directly and indirectly by a major project (Recommendation 2). NSB should also put more emphasis on its long-range planning and priority-setting activities (Recommendation 7) and should periodically reconsider the contribution of every project to agency priorities as part of a more systematic project renewal process (Recommendation 10).

The panel understands that its recommendations cannot guarantee a perfect result or prevent individuals and institutions who are denied awards from complaining about the system. This is especially true of awards for large, one-of-a-kind national facilities that must satisfy many expectations. We believe that the changes recommended in this report will result in a fairer and more understandable process and will increase confidence in, and support by, fair-minded participants and interested groups.

RECOMMENDATIONS

Recommendation 1 : Justification for Major Project Awards
The NSB should ensure that the large-scale research-related projects
that result in major awards are well justified and planned—that is, each
is (a) scientifically justified, (b) technically feasible, (c) designed to
enhance other activities already in place to achieve the proposed project's
goals, (d) of high national priority, and (e) the subject of careful contingency planning.

*Recommendation 2: Involvement and Support of the Research Community in
Planning*
The NSB and NSF should make stronger efforts to see that the basis for
initiating large-scale activities is well explained, understood, and
accepted to the extent possible by affected research communities. NSB
and NSF should take steps to ensure broader consultation with relevant
communities beyond those benefiting directly from a major project
award, including educational, governmental, and industrial organizations and institutions.

Recommendation 3: Primacy of Technical Merit Criteria
The NSB and NSF should continue to make technical excellence the
primary criterion in evaluating the merit of proposals for major awards.
To ensure that research funding is used most effectively, no major
award should ever be made to a project that is not of very high technical
merit. Additional criteria should be used only to choose the best overall
proposal from among those whose technical merit is among the most
highly rated.

*Recommendation 4: Human Resource Development and Equal Opportunity as a
Criterion*
The contribution of every major award proposal to overall human
resource development should be emphasized. The number of students
to be involved—and the inclusion of minorities and women at all levels,
from students to senior investigators and project managers-are important components of human resource development and equal opportunity. They should receive more explicit attention in the review process.

Recommendation 5: Cost Sharing as a Criterion
Cost sharing should be used only to demonstrate commitment to the
project's goals and never simply to extend NSF funds. Where cost sharing is required, NSF should spell out its expectations in the solicitation
announcement. The amount of credit for cost sharing for purposes of

evaluating proposals should be stated clearly and capped at a reasonable level. Due account should be taken of the likelihood that cost-sharing commitments will in fact be met in the out years.

Recommendation 6: A Two-Phase Merit Review Process

For major awards, the peer review part of the merit review process should be conducted in two phases. The first phase would be a strictly technical review; to help assure the primacy of technical merit, only those proposals judged to be technically superior would be forwarded to the second phase for any further consideration. In the second phase, the additional merit criteria would be weighed and balanced with the technical criteria by a more broadly constituted group of reviewers. This second-phase panel would recommend the proposal (or proposals) best meeting the full set of criteria. If the proposal judged to have the highest merit overall is not the one ranked highest in the first phase of review for technical merit, the second-phase panel must explain its recommendation fully. If the top-ranked technical proposal is subsequently not recommended by NSF staff, the chair of the first-phase panel or another member of that panel should present the case for it at the NSB level.

Recommendation 7: Reorienting the NSB Workload

NSB should manage its proposal review workload to ensure that adequate time is left for its most important activities of broad policy direction, long-range planning, and program oversight. That could be accomplished by using its exemption authority more frequently, by raising the delegation threshold, or both.

Recommendation 8: Planning the Review Process and Criteria

NSF and NSB should further strengthen their effort to implement a review process for each major award that (a) imposes a reasonable schedule, (b) identifies the appropriate selection criteria and their relative priority, (c) uses the two-phase review process, (d) selects appropriate reviewers to address each criterion at each stage, and (e) is assisted by a central office for review of major projects that ensures quality and consistency based on extensive experience with such complex project reviews.

Recommendation 9: More and Better Public Documentation of Award Decisions

The review and award process should be fully documented and the results made more accessible than is standard or necessary for traditional individual investigator proposals. This process includes such documentation as site visit and panel reports, and the staff-prepared director's memorandum to the NSB summarizing the review results and recommending the awards. In particular, as recommended above, any decision to pass over the proposal rated highest technically (Phase 1) or to recommend a proposal other than the one selected in Phase 2 of the merit review process must be fully explained, and relevant documents should be publicly available.

Recommendation 10: More Recompetitions
The initial planning of every major award should specify the conditions for renewing, recompeting, or terminating the project. As a general rule, each project (or perhaps, in the case of large national facilities, its management) should be openly recompeted within a time period appropriate to the nature of the activity. Such periodic recompetitions should be preceded by an assessment of whether such an activity, however successful, is still needed or whether the funds would be better used in research areas of higher priority or for other mechanisms (e,g., grants to individual investigators instead of a research center, or a program of university instrumentation awards in place of a central national facility).